Particle Size Measurement
Volume 2

JOIN US ON THE INTERNET VIA WWW, GOPHER, FTP OR EMAIL:

WWW: http://www.thomson.com
GOPHER: gopher.thomson.com
FTP: ftp.thomson.com
EMAIL: findit@kiosk.thomson.com

A service of I(T)P®

Powder Technology Series

EDITED BY

BRIAN SCARLETT and GENJI JIMBO
Delft University of Technology *Chubu Powtech Plaza Lab*
The Netherlands *Japan*

Many materials exist in the form of a disperse system, for example powders, pastes, slurries, emulsions and aerosols. The study of such systems necessarily arises in many technologies but may alternatively be regarded as a separate subject which is concerned with the manufacture, characterization and manipulation of such systems. Chapman & Hall were one of the first publishers to recognize the basic importance of the subject, going on to instigate this series of books. The series does not aspire to define and confine the subject without duplication, but rather to provide a good home for any book which has a contribution to make to the record of both the theory and the application of the subject. We hope that all engineers and scientists who concern themselves with disperse systems will use these books and that those who become expert will contribute further to the series.

Chemistry of Powder Production
Yasuo Arai
Hardback (0 412 39540 1), 292 pages

Particle Size Analysis
Claus Bernhardt
Translated by H. Finken
Hardback (0 412 55880 7), 428 pages

Particle Classification
K. Heiskanen
Hardback (0 412 49300 4), 330 pages

Powder Surface Area and Porosity
S. Lowell and Joan E. Shields
3rd edn, hardback (0 412 39690 4), 256 pages

Pneumatic Conveying of Solids
R.D. Marcus, L.S. Leung, G.E. Klinzing and F. Rizk
Hardback (0 412 21490 3), 592 pages

Principles of Flow in Disperse Systems
O. Molerus
Hardback (0 412 40630 6), 314 pages

Particle Technology
Hans Rumpf
Translated by F.A. Bull
Hardback (0 412 35230 3), 216 pages

Processing of Particulate Solids
J.P.K. Seville, U. Tüzan and R. Clift
Hardback (0 751 40376 8), 384 pages

Particle Size Measurement

Volume 2

Surface area and pore size determination

Fifth edition

TERENCE ALLEN

Formerly Senior Consultant
E.I. Dupont de Nemour and Company
Wilmington, Delaware, USA

CHAPMAN & HALL

London · Weinheim · New York · Tokyo · Melbourne · Madras

Published by Chapman & Hall, 2–6 Boundary Row, London SE1 8HN, UK

Chapman & Hall, 2–6 Boundary Row, London SE1 8HN, UK

Chapman & Hall GmbH, Pappelallee 3, 69469 Weinheim, Germany

Chapman & Hall USA, 115 Fifth Avenue, New York, NY 10003, USA

Chapman & Hall Japan, ITP-Japan, Kyowa Building, 3F, 2-2-1 Hirakawacho, Chiyoda-ku, Tokyo 102, Japan

Chapman & Hall Australia, 102 Dodds Street, South Melbourne, Victoria 3205, Australia

Chapman & Hall India, R. Seshadri, 32 Second Main Road, CIT East, Madras 600 035, India

First edition 1968
Second edition 1975
Third edition 1981
Fourth edition 1990
Fifth edition 1997

© 1968, 1975, 1981, 1990, 1997 T. Allen

Printed in Great Britain at T.J. Press (Padstow) Ltd., Padstow, Cornwall

ISBN 0 412 75330 8
     0 412 75350 2 (2 volume set)

A catalogue record for this book is available from the British Library

Library of Congress Catalog Card number: 96–86676

Contents

Acknowledgements

I would like to express my grateful thanks to Dr Brian H. Kaye for introducing me to the fascinating study of particle characterization. After completing a Masters degree at Nottingham Technical College under his guidance I was fortunate enough to be offered a post at the then Bradford Institute of Technology. At Bradford, Dr John C. Williams always had time for helpful advice and guidance. John became a good friend and, eventually, my PhD supervisor. After more than twenty years at, what eventually became the University of Bradford, I retired from academic life and looked for other interests. It was then I met, once more, Dr Reg Davies who had been a student with me at Nottingham. Reg was working for the DuPont Company who were in need of someone with my background and I was fortunate enough to be offered the position. In my ten years at DuPont I have seen the development of the Particle Science and Technology (PARSAT) Group under Reg's direction. It has been my privilege to have been involved in this development since I consider this group to be pre-eminent in this field. I have learnt a great deal from my thirty or so PARSAT colleagues and particularly from Reg.

My thanks are also due to holders of copyright for permission to publish and to many manufacturers who have given me details of their products.

Terence Allen
Hockessin
DE
USA

Preface to the fifth edition

Particle Size Measurement was first published in 1968 with subsequent editions in 1975, 1981 and 1990. During this time the science has developed considerably making a new format necessary. In order to reduce this edition to a manageable size the sections on sampling dusty gases and atmospheric sampling have been deleted. Further, descriptions of equipment which are no longer widely used have been removed.

The section on dispersing powders in liquids has been reduced and I recommend the book on this topic by my DuPont colleague Dr Ralph Nelson Jr, *Dispersing Powders in Liquids*, (1988) published by Elsevier, and a more recent book by my course co-director at the Center for Professional Advancement in New Jersey, Dr Robert Conley, *Practical Dispersion: A Guide to Understanding and Formulating Slurries*, (1996) published by VCH Publishers.

After making these changes the book was still unwieldy and so it has been separated into two volumes; volume 1 on sampling and particle size measurement and volume 2 on surface area and pore size determination.

My experience has been academic for twenty years followed by industrial for ten years. In my retirement I have been able to utilize the developments in desktop publishing to generate this edition. Although some errors may remain, they have been reduced to a minimum by the sterling work of the staff at Chapman & Hall, to whom I express my grateful thanks.

My blend of experience has led me to accept that accurate data is sometimes a luxury. In developing new products, or relating particle characteristics to end-use performance, accuracy is still necessary but, for process control measurement, reproducibility may be more important.

The investigation of the relationship between particle characteristics to powder properties and behavior is analogous to detective work. It is necessary to determine which data are relevant, analyze them in such a way to isolate important parameters and finally, to present them in such a way to highlight these parameters.

The science of powder technology has long been accepted in European and Japanese universities and its importance is widely recognized in industry. It is my hope that this edition will result in a wider acceptance in other countries, particularly the United States where it is sadly neglected.

Terence Allen

Hockessin, DE 19707, USA

Preface to the first edition

Although man's environment, from the interstellar dust to the earth beneath his feet, is composed to a large extent of finely divided material, his knowledge of the properties of such materials is surprisingly slight. For many years the scientist has accepted that matter may exist as solids, liquids or gases although the dividing line between the states may often be rather blurred; this classification has been upset by powders, which at rest are solids, when aerated may behave as liquids, and when suspended in gases take on some of the properties of gases.

It is now widely recognised that powder technology is a field of study in its own right. The industrial applications of this new science are far reaching. The size of fine particles affects the properties of a powder in many important ways. For example, it determines the setting time of cement, the hiding power of pigments and the activity of chemical catalysts; the taste of food, the potency of drugs and the sintering shrinkage of metallurgical powders are also strongly affected by the size of the particles of which the powder is made up. Particle size measurement is to powder technology as thermometry is to the study of heat and is in the same state of flux as thermometry was in its early days.

Only in the case of a sphere can the size of a particle be completely described by one number. Unfortunately, the particles that the analyst has to measure are rarely spherical and the size range of the particles in any one system may be too wide to be measured with any one measuring device. V.T. Morgan tells us of the Martians who have the task of determining the size of human abodes. Martian homes are spherical and so the Martian who landed in the Arctic had no difficulty in classifying the igloos as hemispherical with measurable diameters. The Martian who landed in North America classified the wigwams as conical with measurable heights and base diameters. The Martian who landed in New York classified the buildings as cuboid with three dimensions mutually perpendicular. The one who landed in London gazed about him dispairingly before committing suicide. One of the purposes of this book is to reduce the possibility of further similar tragedies. The above story illustrates the problems involved in attempting to define the size of particles by one dimension. The only method of measuring more than one dimension is microscopy. However, the mean ratio of significant dimensions for a particulate system may be determined by using two methods of analysis and finding the ratio of the two mean sizes. The proliferation of measuring techniques is due to the wide range of sizes and size dependent properties that have to be measured: a twelve-inch ruler is not a satisfactory tool for measuring mileage or thousandths of an inch and is of limited use for measuring particle volume or surface area. In making a decision on which technique to use, the analyst must first consider the purpose of the analysis. What is generally required is not the size of the particles, but the value of some property of the particles that is size dependent. In such circumstances it is important whenever possible to measure the desired property, rather than to measure the 'size' by some other method and then deduce the required property. For example, in determining the 'size' of boiler ash with a view to predicting atmospheric pollution, the terminal velocity of the particle should be measured: in measuring the 'size' of catalyst particles, the surface area should be determined, since this is the property that determines its reactivity. The cost of the apparatus as well as the ease and the speed with which the analysis can be carried out have then to be considered. The final criteria are that the method shall measure the appropriate property of the particles, with an accuracy sufficient for the particular application at an acceptable cost, in a time that will allow the result to be used.

It is hoped that this book will help the reader to make the best choice of methods. The author aims to present an account of the present state of the methods of measuring particle size; it must be emphasized that there is a considerable amount fo research and development in progress and the subject needs to be kept in constant review. The interest in this field in this country is evidenced by the growth of committees set up to examine particle size measurement techniques. The author is Chairman of the Particle Size Analysis Group of the Society for Analytical Chemistry. Other committees have been set up by The Pharmaceutical Society and by the British Standards Insitution and particle size analysis is within the terms of reference of many other bodies. International Symposia were set up at London, Loughborough and Bradford Universities and it is with the last-named that the author is connected. The book grew from the need for a standard text-book for the Postgraduate School of Powder Technology and is published in the belief that it will be of interest to a far wider audience.

Terence Allen

Postgraduate School of Powder Technology
University of Bradford

Editor's foreword

Particle science and technology is a key component of chemical product and process engineering and in order to achieve the economic goals of the next decade, fundamental understanding of particle processes has to be developed.

In 1993 the US Department of Commerce estimated the impact of particle science and technology to industrial output to be one trillion dollars annually in the United States. One third of this was in chemicals and allied products, another third was in textiles, paper and allied products, cosmetics and pharmaceuticals and the final third in food and beverages, metals, minerals and coal.

It was Hans Rumpf in the 1950s who had the vision of property functions, and who related changes in the functional behavior of most particle processes to be a consequence of changes in the particle size distribution. By measurement and control of the size distribution, one could control product and process behavior.

This book is the most comprehensive text on particle size measurement published to date and expresses the experience of the author gained in over thirty five years of research and consulting in particle technology. Previous editions have already found wide use as teaching and reference texts. For those not conversant with particle size analysis terminology, techniques, and instruments, the book provides basic information from which instrument selection can be made. For those familiar with the field, it provides an update of new instrumentation – particularly on-line or in-process instruments – upon which the control of particle processes is based. For the first time, this edition wisely subdivides size analysis and surface area measurement into two volumes expanding the coverage of each topic but, as in previous editions, the treatise on dispersion is under emphasized. Books by Parfitt[1] or by Nelson[2] should be used in support of this particle size analysis edition.

Overall, the book continues to be the international reference text on the particle size measurement and is a must for practitioners in the field.

Dr Reg Davies

Principal Division Consultant & Research Manager
Particle Science & Technology (PARSAT)
E.I. du Pont de Nemours & Company, Inc.
DE, USA

1   Parfitt, G.D. (1981), *Dispersion of Powders in Liquids, 3rd edn*, Applied Science Publishers, London.
2   Nelson, R.D. (1988), *Dispersing Powders in Liquids, Handbook of Powder Technology Volume 7.* Edited by J.C. Williams and T. Allen, Elsevier.

1

Permeametry and gas diffusion

1.1 Flow of a viscous fluid through a packed bed of powder

The original work on the flow of fluids through packed beds was carried out by Darcy [1], who examined the rate of flow of water from the local fountains through beds of sand of various thicknesses. He showed that the average fluid velocity (u_m) was directly proportional to the driving pressure (Δp) and inversely proportional to the thickness of the bed, L i.e.

$$u_m = K \frac{\Delta p}{L} \qquad (1.1)$$

An equivalent expression for flow through a circular capillary was derived by Hagen [2], and independently by Poiseuille [3], and is known as the Poiseuille equation. The Poiseuille equation relates the pressure drop to the mean velocity of a fluid of viscosity η, flowing in a capillary of circular cross-section and diameter d:
    In deriving this equation it was assumed that the fluid velocity at the capillary walls was zero and that it increased to a maximum at the axis at radius R. The driving force at radius r is given by $\Delta p \pi r^2$ and this is balanced by a shear force of $2\pi r L \eta du/dr$. Hence:

$$2\pi r L \eta du/dr = \Delta p \pi r^2$$

Thus the velocity at radius r is given by:

$$2\eta L \int_0^u du = \Delta p \int_R^r r dr$$

hence the fluid velocity at radius r is given by:

$$u = \frac{\Delta p (r^2 - R^2)}{4\eta L}$$

The total volume flowrate is:

$$\frac{dV}{dt} = 2\pi \int_0^R ru\,dr$$

which on integration gives:

$$\frac{dV}{dt} = \frac{\pi \Delta p R^4}{8\eta L}$$

Dividing by the area available for flow (πR^2) gives the mean velocity in the capillary:

$$\frac{1}{A}\frac{dV}{dt} = u_m = \frac{d^2}{32\eta}\frac{\Delta p}{L} \tag{1.2}$$

It is necessary to use an equivalent diameter (d_E) to relate flowrate with particle surface area for flow through a packed bed of powder [4,5], where:

$$d_E = 4 \times \frac{\text{cross-sectional area normal to flow}}{\text{wetted perimeter}} \tag{1.3}$$

For a circular capillary:

$$d_E = 4 \times \frac{\pi d^2/4}{\pi d} = d$$

Kozeny assumed that the void structure of a bed of powder could be regarded as equivalent to a bundle of parallel circular capillaries with a common equivalent diameter. For a packed bed of powder equation (1.3) may be written:

$$\text{mean equivalent diameter} = 4 \times \frac{\text{volume of voids}}{\text{surface area of voids}}$$

$$d_E = 4 \times \frac{v_v}{S} \tag{1.4}$$

The surface area of the capillary walls is assumed to be equal to the surface area of the powder S. By definition:

$$\text{porosity} = \frac{\text{volume of voids}}{\text{volume of bed}}$$

$$\varepsilon = \frac{v_v}{v_v + v_s}$$

giving:

$$v_v = \left(\frac{\varepsilon}{1-\varepsilon}\right) v_s \qquad (1.5)$$

where v_s is the volume of solids in the bed.
From equations (1.3), (1.4) and (1.5):

$$d_E = 4\left(\frac{\varepsilon}{1-\varepsilon}\right)\frac{v_s}{S} = d$$

Substituting in equation (1.2):

$$u_m = \frac{\varepsilon^2}{(1-\varepsilon)^2}\frac{v_s^2}{S^2}\frac{\Delta p}{2\eta L} \qquad (1.6)$$

It is not possible to measure the fluid velocity in the bed itself, hence the measured velocity is the approach velocity, that is, the volume flow rate divided by the whole cross-sectional area of the bed:

$$u_a = \frac{Q}{A}$$

The average cross-sectional area available for flow inside the bed is εA thus the velocity inside the voids (u_1) is given by:

$$u_1 = \frac{Q}{\varepsilon A}$$

Hence:

$$u_a = \varepsilon u_1 \qquad (1.7)$$

Further, the path of the capillary is tortuous with an average equivalent length L_e, which is greater than the bed thickness L, but it is to be expected that L_e is proportional to L. Thus the velocity of the fluid in the capillary u_m will be greater than u_1 due to the increase in path length:

$$u_m = \left(\frac{L_e}{L}\right)u_1 \tag{1.8}$$

From equations (1.7) and (1.8):

$$u_m = \left(\frac{L_e}{L}\right)\frac{u_a}{\varepsilon} \tag{1.9}$$

Noting also that the pressure drop occurs in a length L_e and not a length L gives, from equations (1.6) and (1.9):

$$u_a = \varepsilon\left(\frac{L}{L_e}\right)\frac{\varepsilon^2}{(1-\varepsilon)^2}\frac{v_s^2}{S^2}\frac{\Delta p}{2\eta L_e}$$

For compressible fluids the velocity u is replaced by $(p_1/\bar{p})u$ where $\bar{p}$ is the mean pressure of the gas in the porous bed and p_1 is the inlet pressure. This correction becomes negligible if Δp is small and p/p_1 is near to unity. Thus:

$$S_w^2 = \frac{1}{k\eta\rho_s^2 u_a}\frac{\varepsilon^3}{(1-\varepsilon)^2}\frac{\Delta p}{L} \tag{1.10}$$

where $S_v = S/v_s$ and $S_v = \rho_s S_w$. S_v is the mass specific surface of the powder and ρ_s is the powder density. In general $k = k_0 k_1$ where $k_1 = (L_e/L)^2$ and, for circular capillaries, $k_0 = 2$. k is called the aspect factor and is normally assumed to equal 5, k_1 is called the tortuosity factor and k_0 is a factor which depends on the shape and size distribution of the cross-sectional areas of the capillaries, hence of the particles which make up the bed.

1.2 The aspect factor k

Carman [6] carried out numerous experiments and found that k was equal to 5 for a wide range of particles. In the above derivation, k_0 was found equal to 2 for monosize circular capillaries. Carman [7] suggested that capillaries in random orientation arrange themselves at a mean angle of 45° to the direction of flow, thus making L_e/L equal to $\sqrt{2}$, k_1 equal to 2 and $k = 5$.

Sullivan and Hertel found experimentally [8] and Fowler and Hertel confirmed theoretically [9] that for spheres $k = 4.5$, for cylinders arranged parallel to flow $k = 3.0$ and for cylinders arranged perpendicular to flow $k = 6$. Muskat and Botsel [cit 7] obtained values of 4.5 to 5.1 for spherical particles and Schriever [cit 7] obtained a value of 5.06. Experimentally, granular particles give k values in the range 4.1 to 5.06.

For capillaries having a Rosin–Rammler distribution of radii [10]:

$$\partial N = K \exp(-r/r_0)\partial r \tag{1.11}$$

where ∂N is the number of capillaries with radii between r and $r + \partial r$ and K and r_0 are constants. The value of S_w derived was greater than that obtained from equation (1.10) by a factor of $\sqrt{3}$, i.e. $k_0 = 2/3$. Thus the permeability equation is not valid if the void space is made up of pores of widely varying radii since the mean equivalent radius is not the correct mean value to be used for the permeability calculation. Large capillaries give disproportionately high rates of flow which swamp the effect of the small capillaries. If the size range is not too great, say less than 2:1, the results should be acceptable. It is nevertheless advisable to grade powders by sieving as a preliminary to surface area determination by permeability and determine the surface of each of the grades independently to find the surface area of the sample. Even if the size range is wide, the method may be acceptable for differentiating between samples. An exception arises in the case of a bimodal distribution; for spheres having a size difference of more than 4:1 the small spheres may be added to large ones by occupying voids. Initially the effect of the resulting fall in porosity is greater than the effect of the decrease in flow rate and the measured surface becomes smaller. When all the voids are filled the value of k falls to its correct value.

Fine dust clinging to larger particles take no part in flow and may give rise to enormous errors. Although the fine dust may comprise by far the larger surface area, the measured surface is the surface of the coarse material. Since the mass of larger particles is reduced because of the addition of fines, the measured specific surface will actually fall.

When aggregates are measured, the voids within the aggregates may contain quiescent fluid, and the measured surface becomes the aggregate envelope surface. It is recommended that high porosities be reduced to a value between 0.4 and 0.5 to reduce this error. In practice this may cause particle fracture, which may lead to high values in the experimental surface area.

The value of k_0 also depends on the shape of the pores [11], lying between 2.0 and 2.5 for most annular and elliptical shapes.

Wasan *et al.* [12,13] discuss the tortuosity effect and define a constriction factor which they include to account for the varying cross-sectional areas of the voids through the bed. They developed several models and derived an empirical equation for regularly shaped particles. The equation is equivalent to replacing the Carman–Kozeny porosity function $\phi(\varepsilon)$ with:

$$\phi(\varepsilon) = 0.2\exp(2.5\varepsilon - 1.6) \text{ for } 0.3 \leq \varepsilon \leq 0.6$$

1.3 Other flow equations

At low fluid velocities through packed beds of powders the laminar flow term predominates, whereas at higher velocities both viscous and kinetic effects are important. Ergun and Orning [14] found that in the transitional region between laminar and turbulent flow, the equation relating pressure gradient and superficial fluid velocity u_f was:

$$\frac{\Delta p}{L} = 150\frac{(1-\varepsilon)^2}{\varepsilon^3}\frac{\eta u_f}{d_{sv}^2} + 1.75\frac{(1-\varepsilon)}{\varepsilon^3}\frac{\rho_s u_f^2}{d_{sv}} \tag{1.12}$$

For Reynolds number less than 2 the second term becomes negligible compared with the first. The resulting equation is similar to the Carman–Kozeny equation with an aspect factor of 25/6. Above a Reynolds number of 2000 the second term predominates and the ratio between pressure gradient and superficial fluid velocity is a linear function of fluid mass flowrate $G = u_f \rho_f$. The constant 1.75 was determined experimentally by plotting $\Delta p/Lu_f$ against G since, at high Reynolds number:

$$\frac{\Delta p}{Lu_f} = \frac{1.75}{d_{sv}}\frac{(1-\varepsilon)}{\varepsilon^3} \tag{1.13}$$

It has been found that some variation between specific surface and porosity occurs. Carman [15] suggested a correction to the porosity function to eliminate this variation. This correction may be written:

$$\frac{(\varepsilon-\varepsilon')^3}{(1-\varepsilon)^2} \text{ for } \frac{\varepsilon^3}{(1-\varepsilon)^2} \tag{1.14}$$

where ε' represents the volume of absorbed fluid that does not take part in the flow. Later Keyes [16] suggested the replacement of ε' by $a(1-\varepsilon)$. The constant a may easily be determined by substituting the above expression into equation (1.28) and plotting $(h_1/h_2)^{1/3}(1-\varepsilon)^{2/3}$ against ε. Neither of these corrections, however, is fully satisfactory.

Harris [17] discussed the role of adsorbed fluid in permeametry but prefers the term 'immobile' fluid. He stated that discrepancies usually attributed to errors in the porosity function or non-uniform packing are, in truth, due to the assumption of incorrect values for ε and S. Associated with the particles is an immobile layer of fluid that does not take part in the flow process. The particles have a true volume v_s and an effective volume v'_s; a true surface S and an effective surface S'; a true density ρ_s and an effective density ρ'_s. The true values can be determined experimentally and, applying equation (1.10), values of S are derived which vary with porosity; usually increasing with decreasing porosity.

Equation (1.10) is assumed correct but the true values are replaced with effective values yielding:

$$\left(\frac{S'}{v'_s}\right)^2 = \frac{\Delta p}{k\eta Lu}\frac{(\varepsilon')^3}{(1-\varepsilon')^2} \tag{1.15}$$

The effective porosity is defined as:

$$\varepsilon' = \left(1 - \frac{w}{\rho'_s AL}\right) = \left(1 - \frac{\rho_B}{\rho'_s}\right) \tag{1.16}$$

where ρ_B, is the bed density.

Combining equations (1.15) and (1.16), the Carman–Kozeny equation take the form:

$$\left(\frac{S'\rho'_s}{v'_s}\right)^2 = \frac{\Delta p}{k\eta Lu}\left[\rho_B\left(\frac{1}{\rho_B} - \frac{1}{\rho'_s}\right)^3\right] \tag{1.17}$$

This equation can be arranged in the form suggested by Carman [7, p. 20]:

$$\left(\frac{\beta}{\rho_B}\right)^{1/3} = \left(\frac{1}{S'_w}\right)^{2/3}\left[\frac{1}{\rho_B} - \frac{1}{\rho'_s}\right] \tag{1.18}$$

where:

$$\beta = \frac{k\eta Lu}{\Delta p}$$

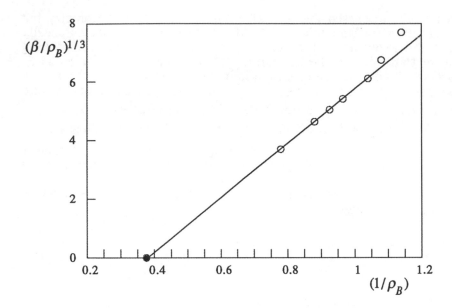

Fig. 1.1 Plot of equation (1.18) for BCR 70 silica; units are SI (kg m s); y-axis multiplied by 10^6; x-axis multiplied by 10^3; i.e. $\rho_B = 1000$ at 1 on the x-axis.

Equation (1.18) expresses a linear relationship between the two experimentally measurable quantities $(\beta / \rho_B)^{1/3}$ and $(1/\rho_B)$. A linear relationship between these quantities means that the effective surface area and mean particle density are constant and can be determined from the slope and the intercept. The fraction of the effective particle volume not occupied by solid material ε_p, apparent particle porosity, is related to density:

$$\varepsilon_p = 1 - \frac{\rho'_s}{\rho_s} \tag{1.19}$$

Schultz [18] examined these equations and found that the effective system surface area was a constant whereas the surface area determined from equation (1.10) varied with porosity. He found that the standard surface area (Blaine number) for SRM 114L, Portland cement, at a porosity of 0.50 agreed well with the Bureau of Standards value of 3380 cm^3 g^{-1}. Further measurements at bed porosities of 0.60 and 0.40 yielded values of 3200 cm^3 g^{-1} and 4000 cm^3 g^{-1} respectively. The effective surface area, which is independent of bed porosity, is

2790 cm^3 g^{-1} at an effective density of 2780 kg m^{-3} as opposed to a true density of 3160 kg m^{-3}. The immobile layer of fluid associated with the bed comprises about 12% of the bed volume ($\varepsilon_p = 0.12$).

An example of this plot is given in Figure 1.1. The graph deviates from linearity at low porosities. For the linear portion $S'_w = 1.134$ m^2 g^{-1} and the intercept on the x-axis yields an effective density ρ'_s of 2661 kg m^{-3} as compared with the quoted density of 2642 kg m^{-3}. The effective surface-volume mean diameter is 2 μm compared to the quoted Stokes diameter of 2.9 μm. For this material, the immobile layer of fluid associated with the bed comprises about 1% of the bed volume.

Replacing ε' by $(v_B - v'_s)/v_B$ yields an alternative form of equation (1.16):

$$\left(\frac{\eta L u v_B}{\Delta p}\right)^{1/3} = \frac{1}{k^{1/3}}\frac{(v_B - v'_s)}{(S')^{2/3}} \tag{1.20}$$

Using the identity $v_B = \left[v_s/(1-\varepsilon)\right]$ equation (1.20) becomes:

$$\left(\frac{\eta L u (1-\varepsilon)^2}{\Delta p}\right)^{1/3} = \left(\frac{\varepsilon - (1 - v_s/v'_s)}{k^{1/3}S'_v\left(v_s/v'_s\right)}\right)^{2/3} \tag{1.21}$$

Harris [17] examined published data which were available in various forms and found that:

- If $(L u \eta/\Delta p)$ is known for a range of ε values then, assuming $k = 5$, equation (1.21) can be used to determine $\left(v_s/v'_s\right)$ and S'_v.
- If it is assumed that the aspect factor (k) varies with porosity, it can be determined as a function of porosity using a previously measured value of S'_v. Equation (1.10) may be written:

$$\left(\frac{\eta L u (1-\varepsilon)^2}{\Delta p}\right)^{1/3} = \frac{\varepsilon^3}{k'S_v^2}$$

Inserting in the left-hand side of equation (1.21) and rearranging gives:

$$\left(\frac{k'}{k}\right) = \left(\frac{S'_v}{S_v}\right)^2 \left(\frac{\varepsilon}{1-(1-\varepsilon)(v'_s/v_s)}\right)^3 \qquad 1.22$$

The ratio (k'/k) can then be calculated for several values of ε and of the parameters (S'_v/S_v) and (v'_s/v_s).

Harris found that most published data gave $(k'/k) = 1.0 \pm 0.10$ and this could be accounted for by modest changes in (S'_v/S_v) and (v'_s/v_s). The tabulated data show that as ε increases (k'/k) decreases at a rate increasing with increasing (v'_s/v_s). Increasing experimental values for k' with decreasing porosity is a common experimental finding. Harris found that the effective volume-specific surface, calculated assuming a constant aspect factor, remained sensibly constant over a range of porosity values.

Equation (1.21) is analogous to equation (1.20) derived from Fowler and Hertel's model, expressed in substantially the same form as Keyes [16], and one developed by Powers [19] for hindered settling, i.e. the equation governing the settling of a bed of powder in a liquid is of the same form as the one governing the flow of a fluid through a fixed bed of powder.

The measured specific surface has been found to decreases with increasing porosity. One way of eliminating this effect, using a constant volume permeameter, is to use equation (1.37) in the form:

$$S_M = \frac{4.93 \times 10^{-7} S_v^2}{1 + 4.93 \times 10^{-7} S_v} \quad S_w = \sqrt{\left(\frac{kt}{\rho_s^2 L}\right) \frac{\varepsilon^{3/2}}{(1-\varepsilon)}}$$

$$S_w = Z \frac{\varepsilon^{3/2}}{(1-\varepsilon)} \sqrt{t}$$

$$\log(t) = 2\log(S_w) - 2\log(Z) - \log\left(\frac{\varepsilon^3}{(1-\varepsilon)^2}\right) \qquad (1.23)$$

Usui [20] replaced the last term with $C + D\varepsilon$ and showed that the relationship between log (t) and ε was linear, proving that C and D were constants independent of ε. A plot of log (t) against ε yields a value for surface area, the calculation being simplified if comparison is made with a standard powder.

Rose [21] proposed that an empirical factor be introduced into the porosity function to eliminate the variation of specific surface with porosity.

$$S'^2 = S^2\left(\frac{1.1}{X} + 140\frac{\varepsilon^3}{(1-\varepsilon)^2}S'\frac{0.2\varepsilon^2}{(1-\varepsilon)^2}\varepsilon^{10}\right) \qquad (1.24)$$

where S' and S are respectively the effective surface and the surface as determined with equation (1.10) and X the porosity function proposed by Carman [6].

1.4 Experimental applications

The determination of specific surface by permeametry was suggested independently by Carman [6] and Dallavalle [22] and elaborated experimentally by Carman [23]. The cement industry published the first test method for the determination of surface area by permeametry, the Lea and Nurse method, and this method is still described in the British Standard [24] on permeametry.
Commercial permeameters can be divided into constant flow rate and constant volume instruments.

The Blaine method [25] is the standard for the cement industry in the United States and, although based on the Carman–Kozeny equation, it is normally used as a comparison method using a powder of known surface area as a standard reference.

The assumptions made in deriving the Carman–Kozeny equation are so sweeping that it cannot be argued that the determined parameter is a surface. First, in many cases, the determined parameter is voidage dependent. The tendency is for the surface to increase with decreasing voidage; low values at high voidage are probably due to channeling i.e. excessive flow through large pores; high values at low voidage could be due to particle fracture or a more uniform pore structure. It thus appears that the Carman–Kozeny equation is only valid over a limited range of voidages. Attempts have been made to modify the equation, usually on the premise that some fluid does not take part in the flow process. The determined surface areas are usually lower than those obtained by other measuring techniques and it is suggested that this is because the measured surface is the envelope surface of the particles. Assuming a stagnant layer of fluid around the particles decreases the measured surface even further.

The equation applies only to monosize capillaries leading to underestimation of the surface if the capillaries are not monosize. Thus the method is only suitable for comparison between similar materials. Because of its simplicity the method is ideally suitable for control purposes on a single product. The method is not suitable for fine powders since, for such powders, the flow is predominantly diffusion.

Permeametry is widely used in the pharmaceutical industry and the technique has been found to give useful information on the assessment of surface area and sphericity of pellitized granules with good agreement with microscopy [26].

1.5 Bed preparation

Constant volume cells and the cell of the Fisher Sub-Sieve Sizer should be filled in one increment only. It is often advantageous to tap or vibrate such cells before compaction but if this is overdone segregation may occur. With other cells the powder should be added in four or five increments, each increment being compacted with the plunger before another increment is added so that the bed is built up in steps. This procedure largely avoids non-uniformity of compaction down the bed, which is likely to occur if the bed is compacted in one operation. To reduce operator bias a standard pressure may be applied (1 MN m^{-2}). In order to test bed uniformity the specific surface should be determined with two different amounts of powder packed to the same porosity. Bed dimensions should be known to within 1%.

1.6 Constant flow rate permeameters

1.6.1 The Lea and Nurse permeameter

In constant flow rate permeameters the flow is maintained constant by using a constant pressure drop across the powder bed. With the Lea and Nurse apparatus [27,28] (Figure 1.2) the powder is compressed to a known porosity ε in a special permeability cell of cross-sectional area A. Air flows through the bed and the pressure drop across it is measured on a manometer as $h_1\rho'g$ and the flow rate by means of a capillary flowmeter, as $h_2\rho'g$ (alternatively a bubble flowmeter can be used). The liquid in both manometers is the same (kerosene or other non-volatile liquid of low density and viscosity) and has a density ρ'. The capillary is designed to ensure that both pressure drops are small, compared with atmospheric pressure, so that compressibility effects are negligible. The bed is formed on a filter paper supported by a perforated plate.

The volume flow rate of air through the flowmeter is given by:

$$Q = \frac{ch_2\rho}{\eta} \tag{1.25}$$

where η is the viscosity of the manometer liquid and c is a constant for a given capillary.

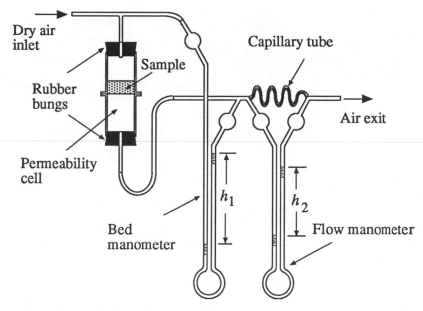

Fig. 1.2 The Lea and Nurse permeability apparatus with manometer and flowmeter.

The pressure drop across the bed as measured by the manometer is:

$$p = h_1 \rho' g \tag{1.26}$$

Substituting equations (1.25) and (1.26) into equation (1.10) gives:

$$S_w = \frac{\sqrt{g/k}}{\rho_s(1-\varepsilon)} \sqrt{\frac{\varepsilon^3 A h_1}{cLh_2}} \tag{1.27}$$

Taking Carman's value of 5 for k this becomes:

$$S_w = \frac{14}{\rho_s(1-\varepsilon)} \sqrt{\frac{\varepsilon^3 A h_1}{cLh_2}} \tag{1.28}$$

Since the terms on the right hand side of the equation are known S_w may be determined.

1.6.2 The Fisher Sub-Sieve Sizer

Gooden and Smith [29] modified the Lea and Nurse apparatus by incorporating a self-calculating chart which enabled surface areas to be read off directly (Figure 1.3). The equation used is a simple transform

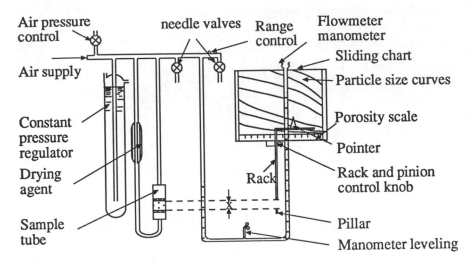

Fig. 1.3 The Fisher Sub-Sieve Sizer.

of the permeameter equation which is developed as follows. The bed porosity may be written:

$$\varepsilon = \frac{v_B - M/\rho_s}{v_B}$$

The volume specific surface may be replaced by the surface-volume mean diameter:

$$\frac{S}{v_s} = \frac{d_s^2}{(\pi/6)d_v^3}$$

$$S_v = \frac{6}{d_{sv}}$$

Also: $\Delta p = (p-f)g; \quad u = fc/A; \quad AL = v_B$

Applying these transformations to equation (1.28) gives:

$$d_{sv} = \frac{60000}{14} \sqrt{\frac{\eta cf\rho_s L^2 M^2}{(v_B\rho_s - M)^3(p-f)}} \tag{1.29}$$

where

d_{sv} = surface-volume mean diameter;

c = flowmeter conductance in mL s^{-1} per unit pressure (g force cm^{-2});

f = pressure difference across flowmeter resistance (g force cm^{-2}).

M = mass of sample in grams;

ρ_s = density of sample in (g cm^{-3});

v_B = bed volume of compacted sample in mL;

p = overall pressure head (g force cm^{-2}).

The instrument chart is calibrated to be used with a standard sample volume of 1cm^3 (i.e. ρ_s grams). It is therefore calibrated according to the equation:

$$d_{sv} = \frac{CL}{(AL-1)^{1.5}} \sqrt{\frac{f}{p-f}} \qquad (1.30)$$

where C is a constant. The chart also indicates the bed porosity ε in accordance with the equation:

$$\varepsilon = 1 - \frac{1}{AL} \qquad (1.31)$$

Since the chart only extends to a porosity of 0.40 it is necessary to use more than ρ_s grams of powder with powders that pack to a lower porosity [30]. If X gram of powder is used, comparison of equations (1.15) and (1.16) shows that the average particle diameter d_{sv} will be related to the indicated diameter d'_{sv} by:

$$d_{sv} = X \left(\frac{1 - X/AL}{1 - 1/AL} \right)^{1.5} d'_{sv} \qquad (1.32)$$

Similarly the bed porosity ε may be calculated from the indicated porosity ε':

$$\varepsilon = 1 - X(1 - \varepsilon') \qquad (1.33)$$

A recommended volume to use in order to extend the range to the minimum porosity is 1.25 cm^3.

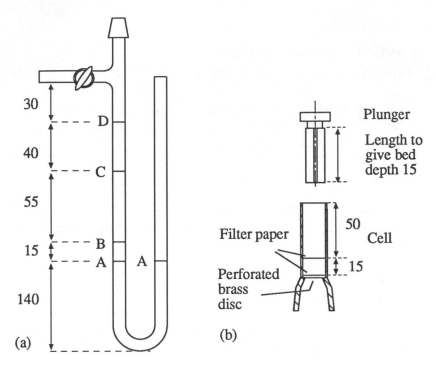

Fig. 1.4 (a) Blaine apparatus and (b) cell and plunger for Blaine apparatus. All dimensions are in millimeters.

The ASTM method for cement standardizes operating conditions by stipulating a porosity of 0.5. This is acceptable since cement is free-flowing and non-cohesive; the range of porosities achievable is therefore limited.

1.7 Constant volume permeameters

1.7.1 The Blaine apparatus

In the apparatus devised by Blaine [31] (Figure 1.4) the inlet end of the bed is open to the atmosphere. Since, in this type of apparatus, the pressure drop varies as the experiment continues, equation (1.10) has to be modified in the following manner. Let the time for the oil level to fall a distance ∂h, when the imbalance is h, be ∂t. Then $\Delta p = h\rho' g$ where ρ' is the density of the oil and:

$$u = \frac{1}{A}\frac{dV}{dt} = \frac{1}{A}\frac{adh}{dt} \tag{1.34}$$

where dV/dt is the rate at which air is displaced by the falling oil. Substituting in equation (1.10) and putting $k = 5$:

$$S_w^2 = \frac{a}{A}\frac{dh}{dt} = \frac{1}{5\eta\rho_s^2}\frac{\varepsilon^3}{(1-\varepsilon)^2}\frac{h\rho'g}{L} \tag{1.35}$$

$$S_w^2 = \int_{h_1}^{h_2}\frac{dh}{h} = \frac{A}{5ah\rho_s^2}\frac{\varepsilon^3}{(1-\varepsilon)^2}\frac{\rho'g}{L} \tag{1.36}$$

$$S_w = \sqrt{\frac{kt\varepsilon^3}{\rho_s^2 L(1-\varepsilon)^2}} \tag{1.37}$$

where k, an instrument constant, is equal to:

$$k = \frac{A\rho'g}{5a\eta\ln(h_2/h_1)} \tag{1.38}$$

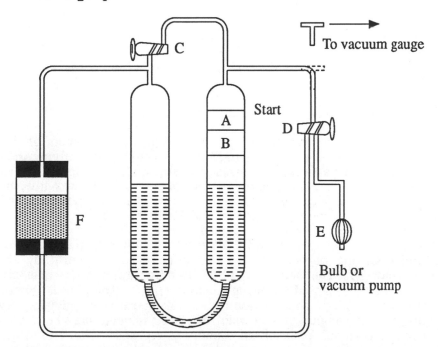

Fig. 1.5 The Griffin Surface Area of Powder Apparatus.

1.7.2 The Griffin Surface Area of Powder Apparatus

A simplified form of the air permeameter was developed by Rigden [32] in which air is caused to flow through a bed of powder by the pressure of oil displaced from equilibrium in two chambers which were connected to the permeability cell and to each other in U-tube fashion. The instrument is available as the Griffin Surface Area of Powder Apparatus (Figure 1.5). The oil is brought to the start position using bulb E with two-way tap C open to the atmosphere. Taps C and D are then rotated so that the oil manometer rebalances by forcing air through the powder bed F. Timing is from start to A for fine powders and start to B for coarse powders.

1.7.3   Reynolds and Branson Autopermeameter

This is another variation of the constant volume apparatus in which air is pumped into the inlet side to unbalance a mercury manometer. The taps are then closed and air flows through the packed bed to atmosphere. On rebalancing, the mercury contacts start–stop probes attached to as timing device. The pressure difference (Δp) between these probes and the mean pressure $\bar{p}$ are instrument constants. The flowrate is given by:

$$\frac{\mathrm{d}v}{\mathrm{d}t} = \frac{1}{p}\frac{\Delta p}{t} \tag{1.39}$$

Substituting this in the Carman–Kozeny equation yields a similar equation to the Rigden equation.

1.7.4   Pechukas and Gage permeameter

This apparatus was designed for the surface area measurement of fine powders in the 0.10 μm to 1.0 μm size range [33]. In deriving their data the inventors failed to correct for slip and, although the inlet pressure was near atmospheric and the outlet pressure was low, no correction was applied for gas compressibility. Their permeameter was modified and automated by Carman and Malherbe [34].

The plug of material is formed in the brass sample tube A (Figure 1.6). Clamp E controls the mercury flow into the graduated cylinder C, the pressure being controlled at atmospheric by the manometer F. The side arm T_1 is used for gases other than air. Calculations are carried out using equation (1.26). The plug is formed in a special press by compression between hardened steel plungers. By taking known weights of a powder, the measurements may be carried out at a known and predetermined porosity, e.g. 0.45. The final stages of compression need to be carried out in small increments and the plungers removed frequently to prevent jamming.

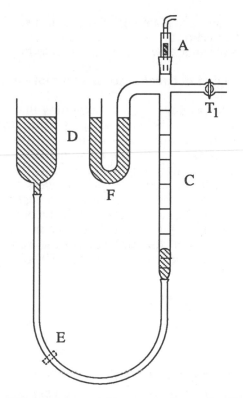

Fig. 1.6 Modified Pechukas and Gage apparatus for fine powders.

1.8 Types of flow

With coarse powders, and pressures near atmospheric, viscous flow predominates and the Carman–Kozeny equation can be used. With compacted beds of very fine powders and gases near atmospheric pressure, or with coarse powders and gases at reduced pressure, the mean free path of the gas molecule is the same order of magnitude as the capillary diameter; this results in slippage at the capillary walls so that the flowrate is higher than that calculated from Poisieulle's premises. If the pressure is reduced further until the mean free path is much larger than capillary diameter, viscosity takes no part in the flow, since molecules collide only with the capillary walls and not with each other. Such free molecular flow is really a process of diffusion and takes place for each constituent of a mixture against its own partial pressure gradient, even if the total pressure at each end of the capillary is the same.

There are therefore three types of flow to consider. In the first the flow is viscous and equation (1.10) may be applied; in the transitional region, in which the mean free path λ of the gas molecules is of the same order as the capillary diameter, a slip term needs to be introduced in order to compensate for the enhanced flow due to molecular diffusion; in the molecular flow region the slip term predominates.

1.9 Transitional region between viscous and molecular flow

Poisieulle's equation was developed by assuming that the velocity at the capillary walls was zero. Rigden [35] assumed that the enhanced volume flowrate due to diffusion may be compensated for by increasing the radius from R to $R + x\lambda$ where $x = (2-f)/f$ and f is the fraction of molecules undergoing diffuse reflection at the capillary walls. Molecules striking smooth capillary walls will rebound at the same angle as the incident angle, i.e. *specular reflection*. The surface of a powder is usually rough and molecules will rebound in any direction, i.e. *diffuse reflection* or inelastic collision. The maximum value for f is unity, which makes $x = 1$ for molecular flow conditions. The flow velocity at a distance r from the center of the capillary becomes (equation 1.2):

$$u = \frac{\Delta p}{32\eta L}\left(\left(R + \frac{2-f}{f}\lambda\right)^2 - r^2\right)$$

Integrating between $r = 0$ and $r = R$, as in the derivation of equation (1.2), and neglecting the term in λ^2, gives the volume flowrate as:

$$\frac{dV}{dt} = \frac{\Delta p\pi R^2}{8\eta L}\left(R^2 + 2R\frac{2-f}{f}\lambda\right)$$

making the mean velocity:

$$\frac{1}{A}\frac{dV}{dt} = u_m = \frac{\Delta p}{32\eta L}\left(d^2 + 2d\frac{2-f}{f}\lambda\right)$$

Making the same substitutions used in deriving equation (1.10) gives:

$$u_a = \frac{\Delta p}{k\eta\Delta\rho_s^2 S_w^2}\frac{\varepsilon^3}{(1-\varepsilon)^2} + \frac{\Delta p}{k\eta L\rho_s S_w}\frac{\varepsilon^2}{(1-\varepsilon)}Z\lambda \qquad (1.40)$$

where $Z = 2x$.

Lea and Nurse [36] arrived at the same equation by assuming a slip velocity at the capillary walls. Carman [7] added an extra term to

Poiseuille's equation, which included a coefficient of external friction, to take account of slip to derive a similar expression.

Alternative forms of equation (1.40) can be found by substituting from the gas equations:

$$\bar{v} = \sqrt{\frac{8RT}{\pi M}} \qquad\qquad \rho_g = \frac{M}{RT}p \qquad\qquad \eta = \frac{1}{2}\rho_g\bar{v}\lambda \qquad (1.41)$$

$$\frac{P_1}{\bar{p}}u = \frac{1}{k\eta\rho_s^2 S_w^2}\frac{\varepsilon^3}{(1-\varepsilon)^2}\frac{\Delta p}{L} + \frac{1}{k\bar{p}\rho_s S_w}\frac{\varepsilon^2}{(1-\varepsilon)}\frac{\Delta p}{L}\sqrt{\frac{2RT}{\pi M}}\frac{8}{3}\partial k_0 \qquad (1.42)$$

Carman and Arnell found $\partial k_0/k = 0.45$ by plotting $(\bar{p}/\Delta p)(V/At)$ against $\bar{p}$ to yield a value $Z = 3.82$. Rigden [36] found an average exp erimental value $Z = 3.80$ but a great deal of scatter was found, i.e. $3.0 < Z < 4.2$.

1.10 Calculation of permeability surface

If the viscous term predominates, the specific surface is determined using the first term of equation (1.40) and, if the compressibility factor is negligible, this takes the form of equation (1.10). When the molecular term predominates, the specific surface is obtained from the second term of equation (1.40). When the two terms are comparable the specific surface is obtained as follows.

The specific surface using the viscous flow term is:

$$S_K^2 = \frac{\Delta p}{5\eta Lu}\frac{\varepsilon^3}{(1-\varepsilon)^2} \qquad (1.43)$$

The specific surface using the molecular flow term is:

$$S_M = \frac{\Delta p}{5\eta Lu}\frac{\varepsilon^2}{(1-\varepsilon)}3.4\lambda \qquad (1.44)$$

Substituting these equations into equation (1.40) gives:

$$\frac{S_K^2}{S_v^2} + \frac{S_M}{S_v} = 1 \qquad (1.45)$$

This is a quadratic in S_v having the following solution:

$$S_v = \frac{S_M}{2} \pm \sqrt{\left(\frac{S_M^2}{4} + S_K^2\right)}$$ (1.46)

Crowl [37] carried out a series of experiments, using pigments, comparing equation (1.40) with Z = 3.4, Rose's equation (1.24) and nitrogen adsorption. He found good agreement between the surfacxe areas determined using equation (1.40) and nitrogen adsorption, a ratio of 0.6 to 0.8 being obtained with a range of surface areas from 1 to 100 m^2 g^{-1}. The areas determined usung Rose's equation were considerably lower, with a ratio ranging from 0.2 to 0.5, being particularly poor with high surface area pigments. With pigments having surface areas above about 10 to 12 m^2 g^{-1} by nitrogen adsorption the agreement was less good but of the same order of fineness as nitrogen adsorption data.

From equations (1.43) and (1.44)

$$S_M = 3.4\left(\frac{1-\varepsilon}{\varepsilon}\right)\lambda S_K^2$$ (1.47)

Using typical values for the variables as an illustration: $\varepsilon = 0.40$ and, for air at atmospheric pressure, $\lambda = 96.6$ nm:

$$S_M = 4.93 \times 10^{-7} S_K^2$$ (1.48)

Substituting in equations (1.29) and (1.30) gives the solutions:

$$S_K^2 = \frac{S_v^2}{1 + 4.93 \times 10^{-7} S_v}$$ (1.49)

$$S_M = \frac{4.93 \times 10^{-7} S_v^2}{1 + 4.93 \times 10^{-7} S_v}$$ (1.50)

Figure 1.7 shows a comparison of the surface areas obtained by using each of the two terms (i.e. S_M and S_K) of the Carman–Arnell equation and the surface obtained by using both terms S_v. The two terms are equal (i.e. $S_M = S_K$) at a surface volume mean diameter of 1.83 μm, each generating 61.8% of the true volume specific surface. At 27 μm, and a porosity of 0.40, the error in assuming that the contribution due to slip is negligible is 5%.

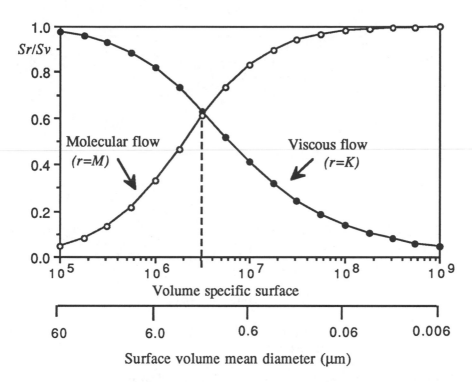

Fig. 1.7 Comparison between the surface area obtained by using each of the two terms of the Carman-Arnell equation (S_K and S_M) and the surface obtained using both terms S_v. The curves represent the fraction of true surface obtained by using the viscous flow term only (black circles) and the slip term only (open circles). Surface area in $m^2 m^{-3}$.

1.11 An experimental approach to the two-term permeametry equation

Allen and Maghafarti [38] used a modified Griffin apparatus to determine the changes in the measured surface area with pressure. They found that the measured permeametry surface (S_K) at atmospheric pressure was porosity dependent and selected the porosity for which this was a maximum for the variable pressure experiment. The volume specific surface (S_V) measured for BCR 70 quartz, determined using the Carman-Arnell equation, remained constant at 3.654 $m^2 m^{-3}$. The powder has a nominal size range of 1.2 to 20 μm and this value of S_V indicates a surface-volume mean diameter of 1.38 μm.

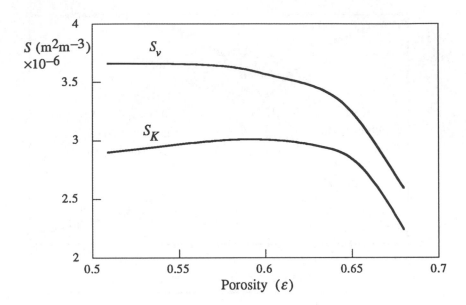

Fig. 1.8 Variation of volume specific surface for BCR 70 quartz using the Carman–Kozeny and the Carman–Arnell equations.

This variation in S_K with porosity is illustrated in Figure 1.8. The low values at high porosity are due to non-homogeneous packing which leads to channeling and enhanced flow through the bed. Several reasons for the low values at higher porosities have been postulated including the presence of an immobile fluid which surrounds the particles and does not take part in the flow process [39]. In reality the fall is due to failure to account for diffusional flow as illustrated in Figure 1.7.

1.12 Diffusional flow for surface area measurement

The rate of transfer of a diffusing substance through unit cross-sectional area is proportional to the diffusion gradient and is given by Fick's laws of diffusion [40]:

$$\frac{1}{A}\frac{dm}{dt} = D\frac{dC}{dx} \tag{1.51}$$

(dm/dt) is the mass flow rate (kg mol s^{-1}) across area A where the concentration gradient is (dC/dx). D, the diffusion constant, has dimensions of m^2 s^{-1} in SI units.

For uni–directional flow into a fixed volume, the increase in concentration with time is given by:

$$\frac{dC}{dt} = D\frac{d^2C}{dx^2}$$

(1.52)

If one face of the powder bed is kept at a constant concentration, i.e. infinite volume source ($C = C_2$ at $x = 0$), while at the other face the initial concentration ($C_1(0)$ at $x = L$, $t = 0$) changes, i.e. fixed volume sink, a finite time will pass before steady state conditions are set up and:

$$C_1 = \frac{\varepsilon AD_s}{LV}\left([C_2 - C_1(0)]t - \frac{L^2}{6D_t}\right)$$

(1.53)

Rewriting in terms of pressure [41]:

$$p_1 = \frac{\varepsilon AD_s}{LV}\left([p_2 - p_1(0)]t - \frac{L^2}{6D_t}\right)$$

(1.54)

where:

p_1 is the (variable) outlet pressure;
p_2 is the (constant) inlet pressure;
$p_1(0)$ is the initial outlet pressure;
V is the outlet volume;
L_e is the equivalent pore length through the bed;
D_t is the unsteady state diffusion constant;
D_s is the steady state diffusion constant.

The two diffusion constants are not necessarily the same. Absorption into pores can take place during the unsteady state period so that the pore volume in the two regimes may be different. Graphs of outlet pressure p_1 against time can be obtained at various fixed inlet pressures p_2. These will be asymptotic to a line of slope:

$$\left(\frac{dp_1}{dt}\right)_{p_2} = \frac{\varepsilon AD}{LV}p_2$$

(1.55)

$$\left[p_1(0) \ll p_2\right]$$

These lines will intersect the line through $p_1(0)$ and parallel to the abscissa at time:

$$t_L = \frac{L_e^2}{6D_t} = \frac{k_1^2 L^2}{6D_t} \tag{1.56}$$

1.13 The relationship between diffusion constant and specific surface

The energy flow rate G through a capillary with a pressure drop across its ends Δp is [42,43]:

$$G = \frac{4}{3} r \sqrt{\left(\frac{2RT}{\pi M}\right)} \frac{A\Delta p}{L} \left(\frac{2-f}{f}\right) \tag{1.57}$$

where R, T and M are the molar gas constant, the absolute temperature and the gas molecular weight and f is the fraction of molecules undergoing diffuse reflection at the capillary walls.

The energy flow rate is related to the diffusion constant by the expression:

$$G = \frac{\varepsilon A\Delta p}{L} D = V \frac{dp}{dt} \tag{1.58}$$

since $p_1 \ll p_2$, $\Delta p = p_2$ and:

$$r = \frac{2}{S_v} \left(\frac{\varepsilon}{1-\varepsilon}\right) \tag{1.59}$$

Combining equations (1.42–1.44) gives, for steady state molecular flow:

$$S_v = 8 \left(\frac{\varepsilon}{1-\varepsilon}\right) \sqrt{\left(\frac{2RT}{\pi M}\right)} \frac{t_L}{L^2} V \left(\frac{dp}{dt}\right)_V = \frac{8}{3} \frac{\varepsilon A\Delta p}{LS_v} \left(\frac{\varepsilon}{1-\varepsilon}\right) \sqrt{\left(\frac{2RT}{\pi M}\right)} \left(\frac{2-f}{f}\right) \tag{1.60}$$

Inserting in equation (1.42) for non-steady-state molecular flow gives:

$$S_v = 16 \left(\frac{\varepsilon}{1-\varepsilon}\right) \sqrt{\left(\frac{2RT}{\pi M}\right)} \frac{t_L}{k_1^2 L^2} \left(\frac{2-f}{f}\right) \tag{1.61}$$

Equation (1.61) is equivalent to equation (1.50) with the constant 3.4 replaced by 8/3 (it being assumed that $f = 1$ for molecular flow).

Derjaguin [44, 45] showed that the constant (4/3) in equation (1.57) should be replaced by 12/13 for inelastic collisions and Pollard and Present [46] use π. Kraus, Ross and Girafalco [47] neglected the tortuosity factor on the grounds that it was already accounted for in the derivation of the diffusion equation. Henrion [48] suggests that molecular diffusion is best interpreted in terms of elastic collisions against the capillary walls.

The general form of equation (1.46) is:

$$\frac{1}{A}\left(\frac{dp}{dt}\right)_V = \beta\left(\frac{\varepsilon^2}{1-\varepsilon}\right)\frac{\Delta p}{LVS_v}\sqrt{\left(\frac{2RT}{\pi M}\right)}$$

(1.62)

The values of β derived by the various researchers are:

Barrer and Grove	8/3	= 2.66
Derjaguin	8/3	= 2.66
Pollard and Present	π	= 3.14
Kraus and Ross	48/13	= 3.70

1.14 Non-steady state diffusional flow

Equation (1.61) was applied by Barrer and Grove [49] with the assumption that $k_1 = \sqrt{2}$ to obtain:

$$S_v = 8\left(\frac{\varepsilon}{1-\varepsilon}\right)\sqrt{\left(\frac{2RT}{\pi M}\right)}\frac{t_L}{L^2}$$

(1.63)

Kraus, Ross and Girafalco assumed no tortuousity factor on the grounds that the internal pore structure is already accounted for and obtained a similar equation to (1.63) with a constant of 144/13. This value was also adopted by Krishnamoorthy *et al.* [50,51].

The apparatus of Kraus *et al.* (Figure 1.9) consists of two reservoirs connected through a cell holding the powder. On the high pressure side the pressure is measured with a mercury manometer and on the low pressure side with a calibrated thermocouple vacuum gauge. The apparatus is first evacuated and flushed with the gas being used. The system is pumped down to 1 or 2cm of mercury and isolated from the vacuum by closing stopcock G. Stopcocks E and F are then closed and the desired inlet pressure established by bleeding gas into reservoir A through tap H. At zero time, stopcock F is opened and the gas allowed to diffuse through the cell C into reservoir B. Figure 1.10 shows a typical flowrate curve. The time lag t_L is determined by extrapolationof the straight line, steady state linear portion of the curve to the initial pressure in the cell and discharge reservoir.

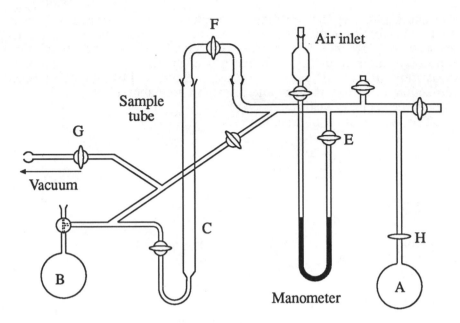

Fig. 1.9 Transient flow apparatus [11].

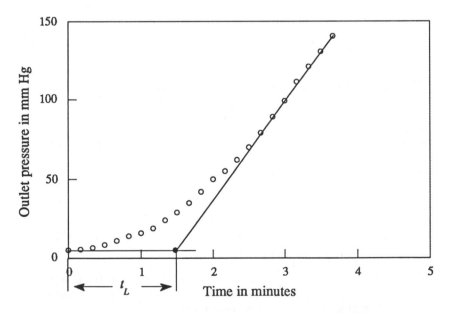

Fig. 1.10 Flow-rate curve for the transient flow apparatus.

This procedure has been used for surface area determination and generates values smaller than those found by gas adsorption. It is difficult to determine t_L accurately, however, and the technique is not recommended for routine analyses.

The example in Figure 1.10 is for rutile titanium dioxide, of density 4260 kg m^{-3}, with a BET surface area of 14.5 m^2 g^{-1}. The time lag of 1.48 min gives a surface area of 6.0 m^2 g^{-1} applying equation (1.47) using the following: $L = 15.3$ cm; $\varepsilon = 0.726$; $T = 20°C$; $R = 8.314 \times 10^3$ J kmol^{-1} K^{-1}; $M = 29.37$ g mol^{-1} for air.

1.15 Steady state diffusional flow

Orr [52] developed an apparatus, which was commercially available from Micromeretics, called the Knudsen flow permeameter which was based on the following form of equation (1.61).

$$S_v = \frac{24}{13}\sqrt{\left(\frac{2}{\pi}\right)}\frac{A\varepsilon^2 \Delta p}{QL\sqrt{(MRT)}} \tag{1.64}$$

where Q is in mol cm^2 s^{-1}.

The flowrate of helium passing through a packed bed of powder is measured together with the upstream pressure p and the pressure drop across the bed Δp. Rearranging equation (1.64) gives the alternative form:

$$S_v = 0.481\frac{A\varepsilon}{q\sqrt{MT}}\left(\frac{\Delta p}{L}\right)\left(\frac{760}{p}\right)\left(\frac{T}{273}\right) \tag{1.65}$$

where q is in cm^2 s^{-1}.

Allen and Stanley-Wood [53–54] developed the surface area meter (Figure 1.11). In this system the inlet pressure p_2 is much greater than the outlet pressure p_1, making $\Delta p = p = p_2$. A graph of energy flow rate $V(dp_1/dt)$ against p_2 gives a straight line from which S_v can be determined. The system is first evacuated with tap 4 closed; taps 1, 2 and 3 are then closed and gas allowed into the inlet side by opening tap 4, thus unbalancing the mercury manometer. Opening tap 1 allows gas to flow through the plug of powder, the flow rate being monitored by the changing inlet pressure which is recorded as a deflection θ on a pen recorder graph. Equation (1.62) is used in the following form:

$$\frac{1}{p}\left(\frac{dp}{dt}\right)_V = \frac{8}{3}\frac{A}{LS_v V}\frac{\varepsilon^2}{1-\varepsilon}\sqrt{\left(\frac{2RT}{\pi M}\right)} = \frac{1}{\theta}\left(\frac{d\theta}{dt}\right) \tag{1.66}$$

For coarse powder it is necessary to correct for the effect of the support plug and filter paper.

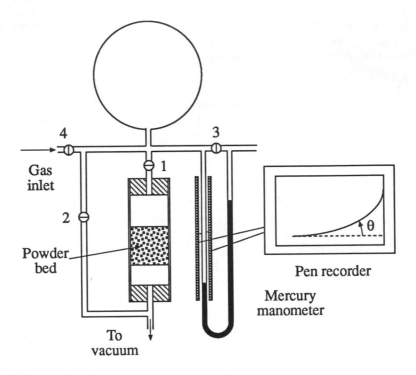

Fig. 1.11 Simple gas diffusion apparatus.

Equation (1.66) may be further simplified if the time is recorded for the pressure to fall from some preset high pressure (p_H) to a preset lower pressure (p_L).

$$S_v = \frac{8}{3} \frac{A^2}{V} \sqrt{\left(\frac{2RT}{\pi M}\right)} \frac{1}{\ln(p_H / p_L)} \frac{\varepsilon^2}{V_p} t \tag{1.67}$$

$$S_v = K \left(\frac{\varepsilon^2}{V_p}\right) t \tag{1.68}$$

Here, the volume of powder in the bed $V_p = AL(1-\varepsilon)$ and K is the product of the instrument constant and the average velocity of the gas molecules (u). For dry air at 273K:

$$u = \sqrt{\frac{8 \times 8.314 \times 273 \times 10^3}{29\pi}}$$

$$= 446 \text{ m s}^{-1}.$$

Under standard operating conditions $A = 5.005$ cm^2, $V = 1000$ cm^3, $p_H = 40$ torr and $p_L = 20$ torr.

Replacing V_p in equation (1.68) with the weight of powder in grams *(w)* gives:

$$S_w = 0.215 \sqrt{\left(\frac{T}{273}\frac{29}{M}\right)\frac{\varepsilon^2 t}{w}} \ \ m^2 g^{-1} \tag{1.69}$$

where T is the operating temperature in absolute units and M the molecular weight of the gas. The consolidating force should be as low as is consistent with a uniform bed; for coarse granular material the bed is loose packed. A timer was incorporated in the commercial version of the instrument (Alstan) which, in one version, also operated as a conventional permeameter [55–57]. Good agreement is found between this instrument and BET gas adsorption for non-porous powders.

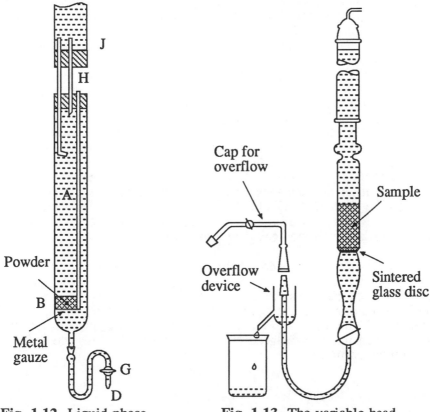

Fig. 1.12 Liquid phase permeameter.

Fig. 1.13 The variable head liquid permeameter.

An instrument based on this design has been examined by Henrion [58], who states that equation (1.69) breaks down for non-random voidage. A classic case is with porous material where diffusion through the wide voids between particles completely swamps the diffusion through the narrow voids within particles. In cases such as this there is good agreement between diffusion and mercury porosimetry.

1.16 The liquid phase permeameter

In the early stages of development of permeametry, liquid permeameters were favored. As long as there is no appreciable mass fraction smaller than 5 µm this technique is still applicable. Below 5 µm the use of liquids becomes unsatisfactory due to settling and segregation, the difficulty of removing air bubbles, aggregation and wetting problems. Gas permeametry was also more attractive due to the higher permeabilities of air and other gases. However the surface areas determined by gas permeametry were less than those determined by liquid permeametry and the difference increased with decreasing size. Though gas permeameters were introduced, they were not placed on a satisfactory basis until the difference between liquid and gas was shown to be due to slip in gases and corrections to the Carman–Kozeny equation derived.

The apparatus used by Carman [6] and others [59–60] is shown in Figure 1.12. The bed is formed by washing a known weight of powder into a uniform tube A, using small increments and allowing each to settle into place with the assistance of gentle suction. The bed rests on a metal gauze B supported by a loosely wound spiral. Liquid flow is adjusted to a steady rate with stopcock G, the difference in level between D and the constant level in A gives the pressure drop causing flow. Air bubbles enter the tube H causing a constant level to be maintained in A and the volume of liquid supplied in a given time is given by the graduated cylinder J.

Dodd, Davis and Pidgeon [61] used the apparatus shown in Figure 1.13 in which the head decreases during the run.

1.17 Application to hindered settling

The settling of particles, constrained to fixed positions, in a stagnant liquid is analogous to the permeametry situation where the liquid is moving and the bed is fixed. For a sedimenting suspension, the pressure head may be replaced by the gravitational minus the buoyant force on the particles:

$$\frac{\Delta p}{L} = (\rho_s - \rho_f)(1 - \varepsilon)g \tag{1.70}$$

Replacing $(\Delta p/L)$ in the permeametry equation (equation 1.10) with the right hand side of this equation, the volume specific surface by $6/d_{sv}$, and eliminating $(\rho_s-\rho_f)g$ using Stokes' equation in the form $(u_{St}=[(\rho_s-\rho_f)gd^2/18\eta])$ gives:

$$u=0.10u_{St}\frac{\varepsilon^3}{(1-\varepsilon)} \tag{1.71}$$

This equation is very similar to the ones derived for the rate of fall of the interface for particles settling in a concentrated suspension with the replacement of d_{sv} with d_{St}.

1.18 Turbo Powdersizer Model TPO-400 In-Line Grain Analyzer

The instrument automatically takes 10–50 kg samples from the discharge from a Nisshin Air Classifier and determines the Blaine number. The instrument was designed as part of a fully automatic system for the cement industry.

1.19 Permoporometry

Commercially available ceramic membranes, with narrow pore size distributions, exhibit properties not shown by polymeric membranes. For example, they can be used at significantly higher temperatures, have better structural stability, can withstand harsher chemical environments, are not subject to microbiological attack and can be backflushed, steam sterilized or autoclaved [62].

Ceramic membranes can be characterized in terms of pore size, pore size distribution, interfacial area, tortuosity, etc. Various tests are carried out to obtain information on the above such as bubble point, SEM, mercury porosimetry, etc. Currently industry uses mercury intrusion porosimetry to characterize pore size distribution. Since mercury cannot differentiate between open and blind pores (closed at one end), mercury porosimetry does not generate the size distribution of pores available for flow. In permoporometry the pores are first filed with a liquid and then the liquid in the pores available for flow is expelled with a second fluid. Since liquid expulsion is unidirectional, this gives an accurate representation of the quality of the filter [63].

1.19.1 Theory

The permeability and separation capability of a membrane can be characterized by the active pore area distribution function. The area of pores with radii in a narrow interval $(r, r + dr)$ can be expressed as:

$$D_2(r) = \frac{dS}{dr}$$

$$\Sigma dS = \Sigma p(r)\, dr \tag{1.72}$$

where S is the active area of all the pores and $p(r)$ is the distribution function. The active area of pores with radii from r_1 to r_2 is given by:

$$S(r_1, r_2) = \int_{r_1}^{r_2} p(r)\, dr \tag{1.73}$$

where

$$\int_{r_{min}}^{r_{max}} p(r)\, dr = 1 \tag{1.74}$$

and r_{min} and r_{max} are the minimum and maximum pore radius, respectively.

The number of pores with radii in the range r to $r + dr$ is:

$$dn = \frac{S p(r)\, dr}{p r^2} \tag{1.75}$$

The flow rate through all the open pores, if they are available for flow, is given by equation 1.2 in the form:

$$\left(\frac{dV}{dt}\right)_{\Delta P} = \frac{S\Delta P}{8\eta L} \sum_1^t r_i^2 p(r_i) = \sum_1^i r_i^2 c_i \frac{dV}{dt} = \frac{S\Delta P}{8\eta L} \int_{r_{min}}^{r_{max}} r^2 p(r)\, dr \tag{1.76}$$

where, from the equation of Young and Laplace;

$$r = \frac{2\gamma \cos(\theta)}{\Delta P} \tag{1.77}$$

An example of a volume flow rate versus pressure difference curve is shown in Figure 1.14. For a pressure difference of less than ΔP_1 the membrane is impermeable. When the pressure reaches ΔP_1 the fluid begins to flow through the biggest pores. With increasing ΔP the liquid is expelled from smaller and smaller pores as they become open or fluid flow. At pressure ∂P all the pores are open and the flow rate becomes proportional to the pressure difference.

Data evaluation proceeds by dividing the pore radii into N discrete intervals where:

$$N\Delta r = r_{max} - r_{min} \tag{1.78}$$

The volume flow rate can be expressed as

$$\left(\frac{dV}{dt}\right)_{\Delta P} = \frac{S\Delta P}{8\eta L}\sum_1^i r_i^2 p(r_i) = \sum_1^i r_i^2 c_i \tag{1.79}$$

where $i = (1,2,........, N)$ so that $r_i = r_{min} - \Delta r(i - \frac{1}{2})$

The parameters c_i can be determined, from experimental data, by linear regression. Further details are available in [64].

The porometer used by Hsieh *et al.* [65]. employed a widely used liquid displacement technique adopted from an ASTM procedure [66], which uses an external air or nitrogen source at pressures up to 1 MPa, allowing pore sizes in the radii range 0.05 to 10 µm [see also BS 3321, BS 1752 (ISO 4793) and BS 5600 (ISO 4003)].

The Coulter Porometer II provides complete pore size flow and number distributions and permeability data in the overall size range 0.05µm to 300µm. Applications include paper, sintered products, porous solids such as core samples, and woven and non-woven fabrics including filter cloths. The instrument operates on the principle that the (gas) pressure required to displace a liquid which freely wets the material is related to the pore radius.

As the pressure increases, liquid will be expelled from pores of radius r at pressure P, where $P = 2\gamma/r$ and γ is the surface tension of the

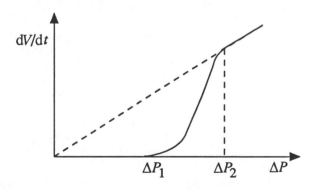

Fig. 1.14 Volume flow rate versus pressure drop through a ceramic membrane.

wetting liquid. The first detectable flow characterizes the largest pore (the bubble point). The pressure is then increased continuously allowing progressively smaller pores to be emptied until the sample is dry. By considering the flowrate of gas through these emptied pores the pore size distribution may be calculated using the Carman–Kozeny equation. Pore size distribution is automatically calculated using a parallel cylindrical pore model as the pressure is gradually increased to 13 bar.

References

1 Darcy, H.P.G. (1856), *Les Fontaines Publiques de la Ville de Dijon*, Victor Dalamont, *1*

2 Hagen, G. (1839), *Ann. Phys. (Pogg. Ann.)*, **46**, 423, *1*

3 Poiseuille, J. (1846), *Inst. de France Acad. des Sci.*, **9**, 433, *1*

4 Blake, F.C. (1922), *Trans. Am. Inst. Chem. Eng.*, **14**, 415, *2*

5 Kozeny, J. (1927), *Ber. Wien. Akad.*, 136A, 271, *2*

6 Carman, P.C. (1938), *J. Soc. Chem. Ing. (Trans.)*, **57**, 225, *4, 11, 32*

7 Carman, P.C. (1956), *Flow of Gases through Porous Media*, Butterworths, *4, 7, 20*

8 Sullivan, R.R. and Hertel, J.L. (1940), *J. Appl. Phys.*, **11**, 761, *5*

9 Fowler, J.L. and Hertel, J.L. (1940), *J. Appl. Phys.*, **11**, 496, *5*

10 Essenhigh, R.H. (1955), *Safety in Mines Res. Est.*, Report No. 120, *5*

11 Orr, C. and Dallavalle, J.M. (1959), *Fine Particle Measurement*, Macmillan, N.Y., *5, 28*

12 Wasan, D.T. *et al.* (1976), *Powder Technol.*, **14**, 209–228, *5*

13 Wasan, D.T. *et al.* (1976), *Powder Technol.*, **14**, 229–244, *5*

14 Ergun, S. and Orning, A.A. (1949), *Ind. Eng. Chem.*, **41**, 1179, *6*

15 Carman, P.C. (1941), ASTM *Symp. New Methods for Particle Size Determination in the Sub–sieve Range*, *24, 6*

16 Keyes, W.F. (1946), *Ind. Eng. Chem.*, **18**, 33, *6, 10*

17 Harris,, C.C. (1977), *Powder Technol.*, **17**, 235–252, *7, 19*

18 Schultz, N.F. (1974), *Int. J. Min. Proc.*, **1**(1), 65–80, *8*

19 Powers, T.C. (1939), *Proc. Am. Concr. Inst.*, **35**, 465, *10*

20 Usui, K. (1964), *J. Soc. Mat. Sci., Japan*, **13**, 828, *10*

21 Rose, H.E. (1952), *J. Appl. Chem.*, **2**, 511, *11*

22 Dallavalle, J.M. (1938), *Chem. Met. Engng.*, **45**, 688, *11*

23 Carman, P.C. (1951), *ASTM, Symp. New Methods for Particle Size Determination in the Sub-SieveRange*, *11*

24 BS 4359: (1982), *Determination of Specific Surface of Powders*, Part 2. Recommended Air Permeability Methods, *11*

25 ASTM C204–68 (1968), *11*

26 Eriksson, M., Nystroem, C. and Alderton, G. (1993), *Int. J. Pharmacy*, **99**(2–3), 197–207, *12*

27 Lea, F.M. and Nurse, R.W. (1949), *Symp. Particle Size Analysis, Trans. Inst. Chem. Eng.*, **25**, 47, *12*
28 Lea, F.M. and Nurse, R.W. (1939), *J. Soc. Chem. Ind.*, **58**, 277, *12*
29 Gooden, E.L. and Smith, C.M. (1940), *Ind. Eng. Chem., Analyt. Ed.*, **12**, 479, *13*
30 Edmondson, I.C. and Toothill, J.P.R. (1963), *Analyst*, October, 805–808, *15*
31 Blaine, R.L. (1943), *ASTM Bulletin* No. 12B, *16*
32 Rigden, P.J. (1947), *J. Soc. Chem. Ind. (Trans.)*, **66**, 191, *18*
33 Pechukas, A. and Gage, F.W. (1946), *Ind. Eng. Chem. Anal. Ed.*, **18**, 370–3, *18*
34 Carman, P.C. and Malherbe, P. le R. (1950), *J. Soc. Chem. Ind.*, **69**, 134, *18*
35 Rigden, P.J. (1954), *Road Res., Tech. Paper* No 28 (HMSO), *20*
36 Lea, F.M. and Nurse, R.W. (1947), S*ymp. Particle Size Analysis, Trans. Inst. Chem. Eng.*, **25**, 47, *20*
37 Crowl, V.T. (1959), *Paint Research Station*, Teddington, Middlesex, Res. Mem., No. 274, **12**, 7, *22*
38 Allen, T. and Maghafarti, R.P. (1981), *Proc. Particle Size Analysis Conf.*, Loughborough, ed. N.G. Stanley–Wood and T. Allen, publ. Wiley, pp. 113–126, *23*
39 Harris, C.C. (1977), *Powder Technol.*, **17**, 235–252, *24*
40 Crank, J. (1946), *Mathematics of Diffusion*, Clarenden Press, Oxford, *24*
41 Babbit, J.D. (1951), *Can. J. Phys.*, **29**, 427, 437, *25*
42 Knudsen, M. (1909), *Ann. Physik.*, **4**(28), 75, 999, *26*
43 Knudsen, M. (1911), *Ann. Physik.*, **4**(34) 25, 593–656, *26*
44 Derjaguin, B. (1946), *C.R. Acad. Sci., USSR*, **53**, 623, *27*
45 Derjaguin, B. (1956), *J. Appl. Chem., USSR*, **29**, 49, *27*
46 Pollard, W.G. and Present, R.D. (1948), *Phys. Rev.*, **73**, 762, *27*
47 Kraus, G., Ross, R.W. and Girafalco, L.A. (1953), *J. Phys. Chem.*, **57**, 330, *27*
48 Henrion, P.N. (1977), *Powder Technol.*, **16**(2), 167–178, *27*
49 Barrer, R.M. and Grove, D.M. (1951), *Trans. Faraday Soc.*, **47**, 826, 837, *27*
50 Krishnamoorthy, T.S. (1966), M.Sc. thesis, Univ. Bradford, *27*
51 Allen, T., Stanley–Wood, N.G. and Krishnamoorthy, T.S. (1966), *Proc. Conf. Particle Size Analysis*, Loughborough, Soc. Anal. Chem., London, *27*
52 Orr, C. (1967), *Anal. Chem.*, **39**, 834, *29*
53 Allen, T. (1971), *Silic. Ind.*, **36**, 718, 173–185, *29*
54 Stanley–Wood, N.G. (1969), PhD thesis, Univ. Bradford, *29*
55 Allen, T. (1978), *Proc. Conf. Particle Size Analysis*, Salford, ed. M.J. Groves, Anal. Div. Chem Soc., publ. Heyden, *31*
56 Stanley–Wood, N.G. (1972), *Proc. Particle Size Analysis Conf.*, Bradford (1970), ed. M.J. Groves and J.L. Wyatt–Sargent, Soc. Analyt. Chem. London, 390–400, *31*

57 Stanley–Wood, N.G. and Chattergee, A. (1974), *Powder Technol.*, **9**(1), 7–14, *31*
58 Henrion, P.N., Greenwen, F. and Leurs, A. (1977), *Powder Technol.*, **16**(2), 167–178, *32*
59 Walther, H. (1943), *Kolloid Z.*, **103**, 233, *32*
60 Wiggins, E.J. Campbell, W.B. and Maas, O. (1948), *Can. J. Res.*, 26A, 128, *32*
61 Dodd, C.G., Davis, J.W. and Pidgeon, F.D. (1951), *Ind. Eng. Chem.*, **55**, 684, *32*
62 Hsieh, H.P. *et al.* (1988), *J. Membr. Sci.*, **39**, 221, *33*
63 Cuperus, F.P. and Smolders, C.A. (1991), *Adv. Colloidal Sci.*, **34**, 135, *33*
64 Mikulásek, P. and Dolecek, P. (1994), *Sep. Sci. Technol.*, **29**(9), 1183–1192, *35*
65 Hsieh, H.P. *et al.* (1988), *J. Membr. Sci.*, **39**, 221, *35*
66 ASTM Procedure No. F316–80, *American Soc. For Testing and Materials,* Philadelphia, Penn., *35*

2

Surface area determination by gas adsorption

2.1 Introduction

When a solid is exposed to a gas, the gas molecules impinging on the surface may not be elastically reflected, but may remain for a finite time. This is designated as adsorption as opposed to absorption, which refers to penetration into the solid body.

The graph of the amount adsorbed (V), at constant temperature, against the adsorption pressure (P), is called the adsorption isotherm. For a gas at a pressure lower than the critical pressure, i.e. a vapor, the relative pressure $x = P/P_0$, where P_0 is the saturation vapor pressure, is preferred.

The amount adsorbed depends upon the nature of the solid (adsorbent), and the pressure at which adsorption takes place. The amount of gas (adsorbate) adsorbed can be found by determining the increase in weight of the solid (gravimetric method) or determining the amount of gas removed from the system due to adsorption by application of the gas laws (volumetric method).

A commonly used method of determining the specific surface of a solid is to deduce the monolayer capacity (V_m) from the isotherm. This is defined as the quantity of adsorbate required to cover the adsorbent with a monolayer of molecules. Usually a second layer may be forming before the monolayer is complete, but V_m is deduced from the isotherm equations irrespective of this. There are also other gas adsorption methods in which the surface area is determined without determining the monolayer capacity.

Adsorption processes may be classified as physical or chemical, depending on the nature of the forces involved. Physical adsorption, also termed van der Waals adsorption, is caused by molecular interaction forces; the formation of a physically adsorbed layer may be likened to the condensation of a vapor to form a liquid. This type of adsorption is only of importance at temperatures below the critical temperature for the gas. Not only is the heat of physical adsorption of the same order of magnitude as that of liquefaction, but physically adsorbed layers behave in many respects like two dimensional liquids.

On the other hand, chemical adsorption (or chemisorption) involves some degree of specific chemical interaction between the adsorbate and the adsorbent and, correspondingly, the energies of adsorption may be quite large and comparable with those of chemical bond formation. Since chemisorption involves chemical bonding, it can occur at temperatures greater than the critical temperature. This implies that chemisorption is restricted to, at most, a single surface layer. With chemisorption, adsorption is limited to specific sites, and the adsorbate molecules have limited ability to migrate about the surface. Thus chemisorption can be used to determine the number of active sites on a surface.

Since physical adsorption is the result of relatively weak interaction between solids and gases, almost all the gas adsorbed can be removed by evacuation at the same temperature at which it was adsorbed. The quantity of physically adsorbed gas at a given pressure increases with decreasing temperature; consequently most adsorption measurements, for the purpose of determining surface area, are carried out at low temperatures. Chemisorbed gas may be difficult to remove merely by reducing the pressure and, when chemisorption does occur, it may be accompanied by chemical changes.

Mathematical theories to describe the adsorption process must, of necessity, be based on simplified models since the shapes of the isotherms depend not only on the specific surface of the powder but also upon the pore structure.

Various boundary conditions limit each of the theories, hence a range of equations have been developed to cover the various phenomena equation developed by Brunauer, Emmett and Teller commonly known as the BET equation. This equation is for multilayer adsorption, but is based upon the Langmuir equation where adsorption is restricted to a monolayer. Both of these equations are developed below, although the application of the Langmuir equation to gas adsorption is restricted to adsorption in micropores where adsorption is limited to a monolayer due to pore geometry. Langmuir adsorption isotherms are common in adsorption of solute from solution.

2.2 Shapes of isotherms

The majority of isotherms may be grouped in the six types shown in Figure 2.1. The first five types were described by Brunauer, Deming, Deming and Teller [1], Type 6 was identified later [2].

The reversible Type 1 isotherm is characterized by a rapid initial amount adsorbed at low pressures followed by a flat region. In some cases the curve is reversible and the amount adsorbed approaches a limiting value. In others, the curve approaches the saturated vapor pressure line (at $x = P/P_0$) asymptotically and the desorption curve may lie above the adsorption curve. For many years it was thought that the shape of the isotherm was due to adsorption being restricted to a monolayer and the isotherm was interpreted on the basis of Langmuir theory [3,4], but it is now generally accepted that the shape is characteristic of micropore filling. This equation therefore has limited applicability to physical adsorption, with wider application to chemisorption and the adsorption of solute from solution. In physical adsorption the restriction to a monolayer may be due to the presence of micropores with pore sizes of a few adsorbate molecular diameters.

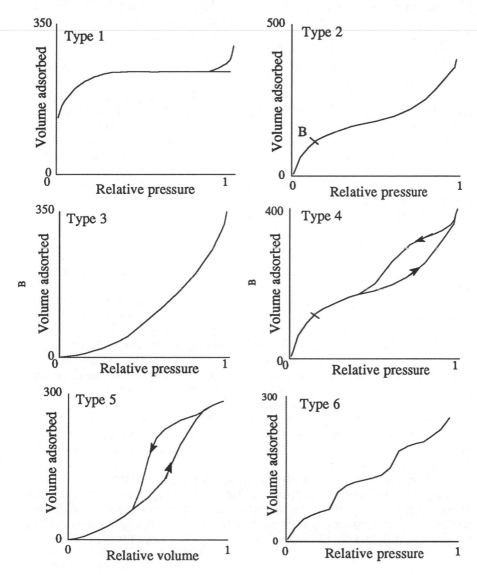

Fig. 2.1 Types of isotherms.

Under these conditions, overlapping pore potentials compress the adsorbate molecules into a smaller volume than they would otherwise adsorbate molecules into a smaller volume than they would otherwise occupy. The concept of surface area becomes meaningless and the limiting amount adsorbed is a measure of micropore volume rather than monolayer surface. The determined volumes will be higher than the true pore volumes, since the adsorbate molecules will be in a condensed liquid state which may approach the volume they would occupy in the solid state. Type 1 isotherms may also occur for adsorption on high energy level surfaces [5].

The reversible Type 2 isotherm is obtained by adsorption on non-porous or macroporous powders and represents unrestricted monolayer–multilayer adsorption on a heterogeneous substrate. Although layers at different layers may exist simultaneously, monolayer completion is assumed at the point of inflection of the isotherm. This is known as point B and was first identified by Emmett and Brunauer [6]. They subsequently developed a theory, containing a constant c, to locate this point. Type 2 isotherms occur for high c values and the 'knee' at the point of inflection becomes more pronounced as the c value increases. Increasing c values indicate increasing affinity between the adsorbate and the adsorbent.

The reversible Type 3 isotherms are convex to the relative pressure axis and exhibits an indistinct point B. Type 3 isotherms arise when the affinity between adsorbate and adsorbent is weaker than the affinity between the adsorbate molecules ($c < 2$). This results in increased adsorption after the interfacial monolayer has formed.

A characteristic feature of Type 4 isotherms is the hysteresis loop. The desorption branch of the isotherm follows a different path to the adsorption branch, although the curve closes as the relative pressure approaches 0.4. This hysteresis is attributed to capillary cracks from which the adsorbate molecules do not desorb as readily as they adsorb, due to vapor lowering over the concave meniscus formed by the condensed liquid in the pores [7,8]. This type of isotherm is found with many mesoporous adsorbents.

Type 5 isotherms result from small adsorbate–adsorbent interaction and are similar to Type 3 isotherms. As with Type 4, the desorption branch of the isotherm differs from the adsorption branch due to the presence of pores.

Type 6 isotherms arise with stepwise multilayer adsorption on a uniform non-porous substrate. The step height represents the monolayer capacity for each adsorbed layer and may remain constant for two or three adsorbed layers. Examples include argon and krypton on graphitized carbon black at liquid nitrogen temperature [9–12].

2.3 Langmuir's equation for monolayer adsorption

The first theoretical equation relating the quantity of adsorbed gas to the equilibrium pressure was proposed by Langmuir [13]. In this model, adsorption is restricted to a monolayer. Under equilibrium conditions, at the interface between the solid and the gas, there is a constant interchange of gas molecules. Langmuir equated the number of molecules evaporating from the surface to the number condensing on to it. Since surface forces are short range, only molecules striking a bare surface are adsorbed; molecules striking a previously adsorbed molecule are elastically reflected back into the gas phase.

From kinetic theory, the number of molecules striking unit area in unit time is given by:

$$Z = \frac{P}{\sqrt{2\pi mkT}} \tag{2.1}$$

k is the Boltzmann constant, m is the mass of the molecule, P is the pressure and T is the absolute temperature.

The number of molecules leaving the surface from unit area in unit time *(n)* depends upon the energy binding the molecules to the surface.

If Q is the energy evolved when a molecule is adsorbed and τ_0 the molecular vibration time, residence time is given by:

$$\tau = \tau_0 \exp\left(\frac{Q}{RT}\right) \tag{2.2}$$

τ_0 is of the order of 10^{-13} s and, for physical adsorption, Q has a value between about 6 and 40 kJ kmol^{-2} [8, p. 463].

The number of molecules evaporating from unit area per second is given by $(1/\tau)$.

If the fraction of the surface covered with adsorbed molecules at pressure P is θ then the rate of adsorption on an area $(1-\theta)$ equals the rate of desorption from an area (θ).

$$\frac{P}{\sqrt{2\pi mkT}} \alpha_0(1-\theta) = \frac{1}{\tau_0} \exp\left(-\frac{Q}{RT}\right) \tag{2.3}$$

where α_0, the condensation coefficient, is the ratio of elastic to total collisions with the bare surface (α_0 tends to unity under conditions of dynamic equilibrium).

If the volume of gas adsorbed at pressure P is V, and the volume required to form a monolayer is V_m, then:

$$\theta = \frac{V}{V_m} \tag{2.4}$$

$$\theta = \frac{bP}{1+bP} \tag{2.5}$$

where:

$$b = \frac{\alpha_0 \tau_0}{\sqrt{2\pi m k T}} \exp\left(\frac{Q}{RT}\right) \tag{2.6}$$

The equation is usually written in the form:

$$\frac{P}{V} = \frac{1}{bV_m} + \frac{P}{V_m} \tag{2.7}$$

A plot of P/V against P (or x/V against x) yields the monolayer capacity V_m and to relate this to surface area it is necessary to know the area occupied by one molecule, σ.

Surface area is calculated from monolayer capacity using the following relationship:

$$S_w = \frac{N\sigma V_m}{M_v} \tag{2.8}$$

where

S_w = mass specific surface ($m^2\,g^{-1}$);
N = Avogadro constant, 6.023×10^{23} molecules mol^{-1};
σ = area occupied by one adsorbate molecule, usually taken as
     $16.2 \times 10^{-20}\,m^2$ for nitrogen at $-195.6°C$;
V_m = monolayer capacity (cm^3 per gram of solid);
M_v = gram molecular volume ($22410\,cm^{-3}\,gmol^{-1}$);
x = relative pressure (P/P_0).

Hence for nitrogen at liquid nitrogen temperature:

$$S_w = \frac{(6.023\times 10^{23})(16.2\times 10^{-20})}{22410} V_m \quad (m^2 g^{-1})$$

$$S_w = 4.35 V_m \quad (m^2\,g^{-1}) \tag{2.9}$$

A basic assumption underlying the Langmuir equation is that the energy of adsorption Q is constant, thus making b constant. This, in turn, implies that the surface is entirely uniform although this is not supported by experimental evidence.

The Langmuir equation has also been derived using thermodynamic [14] and statistical concepts [15].

It is usually assumed in deriving the Langmuir equation that the molecules are adsorbed as wholes (discrete entities) on to definite points of attachment on the surface and each point can accommodate only one adsorbate molecule. If adsorption takes place first on high energy level sites, this must be compensated for by lateral interaction increasing the energy of adsorption of the molecules adsorbed later. Alternatively, if there are no high energy level sites, the energies of the adsorbed molecules are independent of the presence or absence of other adsorbed molecules at neighboring points of attachment.

If, in deriving the Langmuir equation, it is assumed that adsorption is not localized, the rate of condensation is proportional to the total surface and not the bare surface, thus:

$$\frac{P}{\sqrt{2\pi mkT}} \alpha_0 = \left[\frac{1}{\tau_o} \exp(-\frac{Q}{RT}) \right] \theta \tag{2.10}$$

i.e. Henry's Law is obeyed at all pressures.

At high pressures bP is large compared with unity and $V = V_m$, therefore the isotherm approaches saturation.

2.3.1 Henry's law

From equation (2.7), at low pressures bP may be neglected, $(1+bP)$ tends to unity and Henry's law is obeyed [16, p. 104].

$$V = bPV_m$$

$$V = k_H P S_w \tag{2.11}$$

where k_H is Henry's constant. In combination with equation (2.9) this gives:

$$k_H = \frac{bV_m}{S_w} = \frac{b}{4.35} \tag{2.12}$$

Most adsorption equations conform to Henry's law at very low coverage ($x < 0.01$) therefore if Henry's constant is known, the surface area can be determined from a single point on the isotherm.

2.3.2 Halsey equation

If, instead of assuming Q is constant it is assumed that Q is a linear function of the fraction of surface occupied by adsorbate moles θ, the Halsey equation develops [17].

$$mRT \log\left(\frac{P}{P_0}\right) = 1 - \frac{V}{V_m} \qquad (2.13)$$

V is plotted against $\log P$ intersecting the ordinate at $V = V_m$. This equation has been applied to the adsorption of carbon dioxide at $22°C$ on to alumina $(80 < P < 4\,00)$ mmHg and $P_0 = 450$ mm Hg. V was found to equal V_m when $x = 0.10$, the same relative pressure as found for adsorption of carbon dioxide at $-78°C$ and the same value for V_m [18].

2.3.3 Freundlich equation

If it is assumed that Q is a logarithmic function of the fraction of surface occupied by adsorbate moles θ, the Freundlich equation develops:

$$mRT \ln(P/P_0) = \ln(\theta) \qquad (2.14)$$

This has been applied to the adsorption of hydrogen on metallic tungsten [19].

2.3.4 Sips equation

Sips [20] considered a combination of the Langmuir and Freundlich equations:

$$\theta = \frac{AP^{1/n}}{1 + AP^{1/n}} \qquad (2.15)$$

which has the proper limits for monolayer adsorption but reduces to equation (2.13) at low pressures. In a later paper Sips revised his theory [21] and arrived at:

$$\theta = \left(\frac{P}{a+P}\right)^c \qquad (2.16)$$

where a and c are constants.

2.4 BET equation for multilayer adsorption

The most important step in the study of adsorption came with a derivation by Brunauer, Emmett and Teller for the multilayer adsorption of gases on solid surfaces [22]. The multilayer adsorption theory, known generally as the BET equation, has occupied a central position in gas adsorption studies and surface area measurement ever since.

On the assumption that the forces that produce condensation are chiefly responsible for the binding energy of multilayer adsorption, they proceeded to derive an isotherm equation for multilayer adsorption by a method that was a generalization of Langmuir's treatment of the unimolecular layer. The generalization of the ideal localized monolayer treatment is effected by assuming that each first layer adsorbed molecule serves as a site for the adsorption of a molecule into the second layer and so on. Hence, the concept of localization prevails at all layers and forces of mutual interaction are neglected.

S_0, S_1, S_2, S_i represent the fractional surface covered with 0, 1, 2, i layers of adsorbate molecules. At equilibrium, the rate of condensation on S_0 equals the rate of evaporation from S_1 giving:

$$a_1 P S_0 = b_1 S_1 \exp(-Q_1 / RT) \qquad (2.17)$$

where
$$P = \text{pressure ;}$$
$$Q_1 = \text{heat of absorption of the first layer;}$$
$$a_1, b_1 = \text{constants.}$$

$$a_1 = \frac{\alpha_1}{\sqrt{2\pi m k T}}$$

$$b_1 = \frac{1}{\tau_1}$$

it being assumed that the condensation coefficient (α) and the molecular vibration time (τ) vary from layer to layer. This is essentially Langmuir's equation, involving the assumption that a_1, b_1, Q_1 are independent of the number of molecules adsorbed in the first layer. At the first layer, at equilibrium:

$$a_2 P S_1 = b_2 S_2 \exp(-Q_2 / RT) \qquad (2.18)$$

and so on. In general, for equilibrium between the $(i-1)$th and the ith layers

$$a_i P S_{i-1} = b_i S_i \exp(-Q_i / RT))$$ (2.19)

The total surface area of the solid is given by:

$$A = \sum_{i=0}^{\infty} S_i$$ (2.20)

and the total volume of the adsorbate:

$$V = V_0 \sum_{i=0}^{\infty} i S_i$$ (2.21)

where V_0 is the volume of gas adsorbed on unit surface to form a complete monolayer.

Dividing equation (2.21) by equation (2.20) gives:

$$\frac{V}{AV_0} = \frac{V}{V_m} = \frac{\sum\limits_{i=0}^{\infty} i S_i}{\sum\limits_{i=0}^{\infty} S_i}$$ (2.22)

An essentially similar equation had been arrived at earlier by Baly [23], who could proceed further only by empirical means.

Brunauer *et al.* [24] proceeded to solve this summation using two simplifying assumptions, that:

$$Q_2 = Q_3 = Q_4 = \ldots = Q_i = Q_L$$ (2.23)

where Q_L is the heat of liquefaction of the bulk liquid, and:

$$\frac{b_2}{a_2} = \frac{b_3}{a_3} = \ldots = \frac{b_i}{a_i} = g, \text{ a constant}$$ (2.24)

In other words, the evaporation and condensation properties of the molecules in the second and higher adsorbed layers are assumed to be the same as those of the liquid state.

Equation (2.17) can be rewritten:

$$S_1 = y S_0$$ (2.25)

where, from equation (2.17):

$$y = \frac{a_1}{b_1} \exp\left(\frac{Q_1}{RT}\right) \tag{2.26}$$

From equation (2.18):

$$S_2 = XS_1 \tag{2.27}$$

where

$$X = \frac{P}{g} \exp\left(\frac{Q_L}{RT}\right) \tag{2.28}$$

Further:

$$S_3 = XS_2 = X^2S_1 \tag{2.29}$$

and, in the general case for $i > 0$:

$$S_i = XS_{i-1} = X^{i-1}S_1 = yX^{i-1}S_0 = cX^iS_0 \tag{2.30}$$

where

$$c = \frac{y}{X} = \frac{a_1 g}{b_1} \exp\left(\frac{Q_1 - Q_L}{RT}\right) \tag{2.31}$$

and $a_1/b_1 g$ approximates to unity.
Substituting equation (2.30) into equation (2.22):

$$\frac{V}{V_m} = \frac{c \sum_{i=1}^{\infty} iX^i}{1 + c \sum_{i=1}^{\infty} X^i} \tag{2.32}$$

The summation in the denominator is the sum of an infinite geometric progression:

$$\sum_{i=1}^{\infty} X^i = \frac{X}{1-X} \tag{2.33}$$

while that in the numerator is:

$$\sum_{i=1}^{\infty} iX^i = X\frac{d}{dX}\left(\sum_{i=1}^{\infty} X^i\right)\sum_{i=1}^{\infty} X^i = \frac{X}{(1-X)^2}$$

Therefore:

$$\frac{V}{V_m} = \frac{cX}{(1-X)(1-X+cX)} \tag{2.34}$$

On a free surface the amount adsorbed at STP is infinite. Thus at $P=P_0$, the saturation vapor pressure of the adsorbate at the temperature of adsorption, x is equal to 1, making $X = 1$, in order to make $V= \infty$. Therefore, substituting $X= 1$ and $P = P_0$ in equation (2.28) and dividing the result by the original equation (2.28) gives:

$$X =(P/P_0) \tag{2.35}$$

i.e. $X = x$

Substituting in equation (2.34):

$$V = \frac{cPV_m}{(P_0 - P)\left[1+(c-1)\dfrac{P}{P_0}\right]} \tag{2.36}$$

which transforms to:

$$\frac{P}{V(P_0-P)} = \frac{1}{cV_m} + \frac{c-1}{cV_m}\frac{P}{P_0}$$

This may be written:

$$\frac{x}{V(1-x)} = \frac{1}{cV_m} + \frac{c-1}{cV_m}x \tag{2.37}$$

which is commonly known as the BET equation named after the original formulators Brunauer, Emmett and Teller. A plot of $P/[V(P_0-P)]$ against (P/P_0) yields a line of slope $[(c-1)/cV_m]$ and intercept $[1/cV_m]$.

This equation is capable of describing Type 1, Type 2 and Type 3 isotherms, depending on the value of the constant c. It is found that only Type 2 isotherms (i.e. those with high c values) have well-defined knee bends which are essential for accurate V_m values. The preference for using nitrogen at liquid nitrogen temperatures is due to the fact that, with all solids so far reported, this gas exhibits higher c values than alternative gases.

For Type 2 isotherms, the BET equation has been found to hold between 0.05 and 0.35 relative pressure, but examples have been reported where this range has been extended or shortened [25].

The internal consistency of the BET method has been demonstrated by many authors [7,16] by their measurements on several solids. The degree of correspondence between the specific surfaces obtained with several adsorbates allows confidence to be placed in the method.

The intercepts obtained are usually very small. Negative values have also been reported [26]. MacIver and Emmett [27] found that this could be accounted for by the BET equation not fitting the experimental data for $x > 0.2$.

2.4.1 Single-point BET method

When $c \gg 1$ equation (2.37) takes the form:

$$(1-x)V = V'_m \tag{2.38}$$

Hence, it may be assumed that for high c values the BET plot passes through the origin and the slope is inversely proportional to the monolayer capacity. Thus only one experimental point is required. This simplification is often used for routine analyses. The error in V_m that results from using "he single point technique depends on the relative pressure and the value of c and may be determined from equations (2.37) and (2.38):

$$\text{Error} = \frac{V_m - V'_m}{V_m}$$

$$\text{Error} = 1 - \frac{cx}{1+(c-1)x} \tag{2.39}$$

For $c = 100$, and a single point measurement at $x = 0.1$, 0.2 and 0.3 respectively, the errors are 8%, 4% and 2%.

A fixed relative pressure of either 0.2 or 0.3 is often used with the single point BET method. The resulting errors are presented in Table 2.1 and it can be seen that this method should not be used for materials having c values less than 100.

Table 2.1 Error in calculated surface area through using the single
point BET method

c	$x = 0.2$	$x = 0.3$
3	57.14	43.75
5	44.44	31.82
10	28.57	18.92
30	11.77	7.22
100	3.85	2.28
1000	0.40	0.23

2.4.2 Discussion of the BET equation

The BET equation includes isotherms of types 1, 2 and 3 but not types
4 and 5. Brunauer *et al.* [28] derived a new isotherm equation to cover
all five types. The BET equation has also been derived by statistical
reasoning by several authors [16].

The BET model has been criticized on the grounds that, although c
is assumed constant, it is known that the first molecules to be adsorbed
generate more energy than subsequent molecules. For c to be a
constant it is necessary that the surface be energetically homogeneous.
Experimental measurements of variation of heat of adsorption with
coverage show that the first molecules to be adsorbed generate more
energy than subsequent molecules. Because of this, the theory is not
applicable at relative pressures less than about 0.05.

A further criticism is that horizontal interactions are neglected, i.e.
only the forces between the adsorbing molecule and the surface are
considered. Thus the first molecule to be adsorbed is considered to
generate the same energy as the final molecule to fill a monolayer,
although in the former case the molecule has no near neighbors
whereas in the latter it has six. These two effects must, in part, cancel
each other out.

It is also questionable whether adsorbing molecules in the second
and subsequent layers should be treated as being equal. One would not
expect a sudden transition in the energy generated between the first and
subsequent layers.

Cassel [29,30] showed, using Gibbs' adsorption isotherm, that the
surface tension of the adsorbed film at $P = P_0$ is negative, arising from
the total disregard of the interaction forces. Since the BET model
assumes the existence of localized adsorption at all levels, the molecules
being located on top of one another, and since the adsorption can take
place in the nth layer before the $(n-1)$th layer is filled, the adsorbed
phase is built up not as a series of continuous layers, but as a random
system of vertical molecular columns. Halsey [31] pointed out that the
combinational entropy term associated with these random molecular
piles is responsible for the stability of the BET adsorbed layers at

pressures below the saturation pressure. This large entropy term is the cause of too large adsorptions observed when $x > 0.35$.

Gregg and Jacobs [32] doubted the validity of the assumption that the adsorbed phase is liquid-like, and found that the integration constant in the Clausius-Clapeyron equation, as applied to adsorption and vapor pressure, do not show the inter-relationship demanded by the BET theory. They conclude that any constant can be used in the place of P_0 and that the correspondence between the adsorbed and liquid phase is a loose one, arising out of the fact that the same type of force is involved.

Halsey [33] pointed out that the hypothesis that an isolated adsorbed molecule can adsorb a second molecule on top, yielding the full energy of liquefaction, and that in turn the second molecule can adsorb a third, and so on, is untenable. If the molecules are hexagonally packed, one would be much more likely to find a second layer molecule adsorbed above the center of a triangular array of first layer molecules. Applying this modification, the BET model results in very little second layer adsorption when the first layer is one third full and virtually no adsorption in the third layer except at high relative pressures. Values of V_m given by this modified theory do not differ appreciably from those given by the simple BET theory.

The monolayer capacity occurs at the so-called point B on the $V–P$ isotherm where the slope of the isotherm changes on completion of a monolayer. This is normally situated within the relative pressure region $0.05 < x < 0.15$. The BET theory predicts a point of inflection at variable coverage depending on the c value.

At relative pressures greater than about 0.35 the BET equation predicts adsorption greater than observed and the equation breaks down. A critical review of the limitations of the BET theory has been presented by Dollimore *et al.* [34].

2.4.3 Mathematical nature of the BET equation

The BET equation is equivalent to the difference between the upper branches of two rectangular hyperbolas [35] and may be written:

$$\frac{V}{V_m} = \frac{1}{1-x} - \frac{1}{1+(c-1)x} \tag{2.40}$$

For c less than 2 and positive, a type 3 isotherm is generated. Values of c less than unity infer that the cohesive force between the adsorbate molecules is stronger than the adhesive force between the adsorbate and adsorbent.

The point of inflection of the isotherm may be determined by differentiating twice with respect to x and equating to zero. This gives:

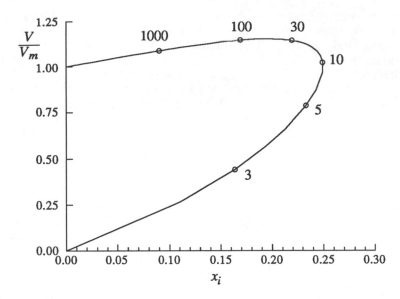

Fig. 2.2 Variation of the location of the point of inflection of the BET isotherm with c value.

$$x_i = \frac{(c-1)^{1/3} - 1}{(c-1)^{2/3}} \tag{2.41}$$

$$\left(\frac{V}{V_m}\right)_i = \frac{(c-1)^{2/3} - 1}{(c-1)^{2/3} - (c-1)^{1/3} + 1} \tag{2.42}$$

By plotting $\left(V/V_m\right)_i$ against x_i for variable c it can be seen (Figure 2.2) that at high c values, monolayer adsorption occurs at or near the point of inflection. For low c values the ratio of V to V_m at the point of inflection becomes progressively smaller until for $c < 2$ the point occurs at negative values of x. e.g. for a c value of 100 the point of inflection occurs at a relative pressure $x_i = 0.169$ when the volume coverage $V = 1.148V_m$.

Genot [36] described a more accurate method for determining monolayer capacity by considering the mathematical nature of the BET equation. The treatment is as follows. At point M on the isotherm let $V = V_{m}$, $P = P_m$ and $x = x_m$. Then:

$$x_m = \frac{1}{1 + \sqrt{c}}$$

At $V = 0.5V_m$, let $P = P_{0.5}$ and $x = x_{0.5}$ then:

$$c = \frac{(1+x_{0.5})^2}{x_{0.5}(1+x_{0.5})}$$

For high c values:

$$x_{0.5} = \frac{1}{c+3} \approx \frac{1}{c} = \frac{\tau_2}{\tau_1}$$

the inverse ratio of the lifetimes of the adsorbed molecules in the first and subsequent layers. The tangent at point M passes through the point $G(x = 1, V/V_m = 3)$ for all c since:

$$\left(\frac{d(V/V_m)}{dx}\right)_M = \left(\frac{d(\ln V)}{dx}\right)_M = \frac{2}{1-x_m} = -2\left(\frac{d(\ln(1-x))}{dx}\right)_M$$

or

$$\left(\frac{d(\ln V)}{(d\ln(1-x))}\right)_M = \left(\frac{d(\log V)}{(d\log(1-x))}\right)_M = -2$$

Hence, point M will result from the determination of the tangent of slope -2 on the graph of $\log(V)$ against $\log(1-x)$. This treatment produces more standardized results than using the conventional BET plot.

Hill [37] showed that when sufficient adsorption had occurred to cover the surface with a monolayer, some fraction of the surface $(\theta_0)_m$ remains bare. Hill established that this fraction is related to the c value:

$$(\theta_0)_m = \frac{c^{0.5}-1}{c-1} \tag{2.43}$$

Lowell [38] extended the argument to show that the fraction of surface covered by molecules i layers deep is, for $i = 1, 2, 3,....$:

$$(\theta_i)_m = c\left(\frac{c^{0.5}-1}{c-1}\right)^{i+1} \tag{2.44}$$

For example, for $c = 100$:

i	0	1	2	3	4	5
$(\theta_i)_m$	0	0.8264	0.0751	0.0068	0.0006	0.00006

i.e. 82.64% of the surface is covered with a monolayer and 9.09% is bare. For the special case when $c = 1$:

$$(\theta_i)_m = \lim_{c \to 1} \left(\frac{c^{0.5} - 1}{c - 1} \right) = 0.5$$

Lowell and Shields [39] also showed that, when the BET equation is solved for the relative pressure corresponding to monolayer coverage, this pressure is also the fraction of the surface unoccupied by adsorbate.

2.4.4 *n-layer BET equation*

If, owing to special considerations, the number of layers cannot exceed n, the BET equation becomes:

$$\frac{V}{V_m} = \frac{cx}{(1-x)} \frac{1 - (n+1)x^n + nx^{n+1}}{1 + (c-1)x - cx^{n+1}} \tag{2.45}$$

This equation applies to adsorption in a limited space such as a capillary. When $n = 1$ it reduces, at all values of c, to the Langmuir equation (unlike the simple BET equation which only reduces when $x \ll 1$ and $c \gg 1$). It also reduces to the same form as the BET equation when $n = 2$ and $c = 4$ [40], and for $n \gg 3$ it is capable of reproducing the shape of all five isotherm types provided c lies within certain narrow limits [41] or x lies between 0.05 and 0,35 [42]. Brunauer, Emmett and Teller successfully applied this isotherm to a variety of isotherms obtained by themselves and others.

The n-layer equation may be written:

$$\frac{V}{V_m} = \frac{c\phi(n,x)}{(1 + c\theta(n,x))} \tag{2.46}$$

where:

$$\phi(n,x) = \frac{x(1 - x^n) - nx^n(1 - x)}{(1-x)^2}$$

and

$$\theta(n,x) = \frac{x(1 - x^n)}{(1-x)}$$

Joyner *et al.* [43] compiled tables of ϕ and for θ for increasing x. Using these tables the best straight line is selected for ϕ against θ since the linear form of the equation is:

$$\frac{\phi}{V} = \frac{1}{cV_m} + \frac{\theta}{V_m} \qquad (2.47)$$

The BET plot of $[x/(1-x)](1/V)$ against x is convex to the x-axis for microporous materials since, as the pressure is increased, the increments adsorbed are smaller than for non-porous materials due to micropore filling. The n-layer equation will produce a straight line with an increased intercept on the ϕ axis (i.e. a lower c value) and a higher specific surface. Low c values are improbable with microporous materials, hence the validity of the technique is questionable. Gregg and Sing [44] go so far as to state that n is no more than an empirical parameter adjusted to give best fit to the experimental data. It would seem doubtful if the procedure possesses any real advantage over the more conventional BET method.

2.4.5 Pickett's equation

Pickett [45] modified the n-layer BET treatment to take into account the decrease in probability of escape from an elemental area covered with n layers as adjacent elements also become covered with n layers. This leads to the result that there can be no evaporation from such an area. With this assumption Pickett derived the equation:

$$\frac{V}{V_m} = \frac{cx(1-x^n)}{(1-kx)(1+(c-1)kx)} \qquad (2.48)$$

2.4.6 Anderson's equation

Anderson [46] made the assumption that the heat of adsorption in the second to about the ninth layer differs from the heat of liquefaction by a constant amount. The resultant equation is similar to the BET equation but contains an additional parameter k ($k < 1$).

$$\frac{V}{V_m} = \frac{kcx}{(1-kx)(1+(c-1)kx)} \qquad (2.49)$$

A further modification took into account the decrease in area available in successive layers, a situation likely to prevail with porous solids [47,48]. This leads to the equation:

$$\frac{V}{V_m} = \frac{kcx}{(1-ikx)(1+(c-i)kx)} \tag{2.50}$$

where $i = A_n/A_{n-1}$, A_n is the number of molecules required to fill the nth layer and A_{n-1} is the number of molecules required to fill the $(n-1)$th layer.

Brunauer *et al.* [49] introduced a parameter K into the BET equation where K is a measure of the attractive field force of the adsorbent. Although the derived equation is identical to Anderson's the model on which it is based is different.

Barrer *et al.* [50] derived eighteen analogs of the BET equation by making various assumptions as to the evaporation–condensation properties of the molecules in each layer.

2. 5 Polanyi's Theory for micropore adsorption

Microporous materials exhibit type 1 isotherms since the size of the pores restricts adsorption into a few layers. The field strength within the pores is so great that it is difficult to determine whether the adsorbate packs as a liquid or in a more condensed form. Polanyi [51] assumed that above the critical temperature the adsorbate can adsorb only as a vapor whose density increases as it approaches the surface; around the critical temperature the vapor near the surface starts to liquefy, and substantially below the critical temperature the adsorbate completely liquefies. Under this final condition, the adsorbed volume of liquid (v) can be determined from the adsorption isotherm. Polanyi potential theory states that the adsorption potential for adsorbate in the liquid state is given by the isothermal work required (ε) to compress the vapor from its equilibrium pressure (P) to its saturated vapor pressure (P_0).

$$\varepsilon = RT \ln\left(\frac{P}{P_0}\right) \tag{2.51}$$

A plot of ε versus v is called a characteristic curve.

Niemark [52] stated that the Polanyi theory of adsorption was gaining widespread acceptance due to the clarity of the underlying assumptions and the ensuing thermodynamic relationships.

2.6 Dubinen–Radushkevich equation

This equation is based on the Polanyi potential theory for the determination of micropore volume [53,54]. In this theory it is postulated that the force of attraction at any point in the adsorbed film is given by the adsorption potential (ε), defined as the work done by the adsorption forces to bring a molecule from the gas phase to this point.

If two adsorbates fill the same adsorption volume their adsorption potentials will differ only because of differences in their molecular properties. The ratio of adsorption potentials was assumed by Dubinen [55] to be a constant which he defined as an 'affinity constant (β)' since it is a measure of the relative affinities of different molecules for a surface. Using benzene as a reference vapor the affinity ratios for other vapors were determined [56].

$$\beta = \frac{\varepsilon_1}{\varepsilon_2} \qquad (2.52)$$

where the suffixes refer to the respective vapors. If suffix 2 is taken as an arbitrary standard then $(\varepsilon_1/\varepsilon_0) = \beta_0$. Assuming that the adsorption volume may be expressed as a Gaussian function of the corresponding adsorption potential then, for the standard vapor:

$$\frac{V}{\rho V_p} = \exp\left(-k\varepsilon_0^2\right) \qquad (2.53)$$

where V is the volume of vapor adsorbed at STP, ρ is the density of the liquid adsorbate, V_p is the micropore volume and k is a constant.

The equation can be written in the linear form, known as the Dubinen-Radushkevich (D-R) equation:

$$\log W = \log(v_0\rho) - D\left[\log\left(\frac{P_0}{P}\right)\right]^2 \qquad (2.54)$$

where $D = 2.303K(RT/\beta)^2$, W and ρ are the weight adsorbed and the liquid adsorbate density, K is a constant determined by the shape of the pore size distribution curve and v_0 is the total micropore volume.

A plot of log W against $\left[\log(P_0/P)\right]^2$ should give a straight line with an intercept from which the micropore volume (v_0) can be calculated.

The equation has been found to apply for the adsorption of nitrogen, saturated hydrocarbons, benzene and cyclohexane over the relative pressure range 10^{-5} to 10^{-1} on a range of microporous samples [57]. However, the equation needs correction for adsorbate compressibility [58,59]. It has also been used for determining the micropore volume of carbons [60–66] and the thermodynamic parameters of adsorption [67–70].

Many isotherms of vapors on non-porous solids can be linearized in the sub-monolayer region by applying this equation [71–73]. Rosai and Giorgi [74] applied it to the adsorption of argon and krypton on to barium at 77.3K and state that this enables one to distinguish between barium films having different surface characteristics and this gives a qualitative description of the dynamics of sintering.

Stoekli [75] used a more generalized form of the D–R equation for the filling of a heterogeneous micropore system and found that the constant k decreased with increasing $(RT/\beta)^2 \log(P_0/P)$.

Marsh and Rand critically appraise the D–R equation and state that it predicts a Raleigh distribution of adsorption free energy, and only when this distribution is present in microporous solids will a completely linear plot result. Adsorption of carbon dioxide, nitrogen and argon at 77 K on various microporous carbons were examined and in no case was the complete Raleigh distribution found to apply. In order to obtain meaningful parameters they recommended that the experimental data be extended to as near unit relative pressure as possible [76,77].

2.6.1 Dubinen–Askakhov equation

Rand [78] showed that the D-R data could be made more linear by using the more general Dubinen–Askakhov (D–A) equation.

$$\frac{V}{\rho V_p} = \exp\left[-k\left(\frac{\varepsilon}{E}\right)^n\right]$$

(2.55)

The significance of this equation is discussed and it is shown that, for carbon dioxide and nitrogen adsorption on microporous carbons, the constants E and k are independent of the physical nature of the adsorbent and the temperature of adsorption. The energetic heterogeneity of the surface is therefore described by one parameter n.

The adsorption of carbon dioxide at 273 K is frequently used to determine V_p and, since at this temperature pressurized equipment is needed to reach P_0, it has been suggested that adsorption should be measured at pressures less than one atmosphere and extrapolated using the D-A equation [79]. In some cases a change of slope occurs at $\varepsilon=10$ kJ mol^{-1} and at one atmosphere these plots cannot be treated this way. It was concluded that the D-A equation is empirical and should be used with caution when the V-t method cannot be used.

2.6.2 Dubinen-Kaganer equation

Kaganer [80,81] modified Dubinin's method in order to calculate the surface area within micropores where, essentially, he assumed that (v_0) was the volume adsorbed at monolayer coverage. He assumed a

Gaussian distribution of adsorption over the surface sites i.e. he assumed $V/V_m = \exp(-k\varepsilon^3)$.

Hence the D–R equation becomes:

$$\ln\left(\frac{V}{V_m}\right) = -D\left(\frac{\ln P_0}{P}\right)^2 \qquad (2.56)$$

The D–K equation has the same form as the D–R equation except that the fractional surface coverage (V/V_m) replaces the fractional volume filling of pores. The equation is applicable to relative pressures below 0.01. Kaganer examined a range of powders, using nitrogen as adsorbate at relative pressures between 0.0001 and 0.01, and obtained good agreement with BET.

Ramsay and Avery [82] investigated nitrogen adsorption in microporous silica compacts for $10^{-2} > x > 10^{-4}$ and found that $\varepsilon < 7$ kJ mol^{-1} for loose powder and increased with compaction and decrease in pore size. They found close agreement between amount adsorbed at monolayer coverage and intercept volume using the D–K equation, which indicated surface coverage as opposed to volume filling.

2.7 Jovanovic equations

Jovanovic [83] derived two equations for monolayer and multilayer adsorption.

$$V_m = V(1 - \exp(-ax)) \qquad (2.57)$$

$$V_m \exp(bx) = V(1 - \exp(-ax)) \qquad (2.58)$$

where a and b are constants which describe adsorption in the first and subsequent layers respectively.

$$a = \frac{\sigma\tau P_0}{\sqrt{2umkT}} \qquad (2.59)$$

$$b = \frac{\sigma\tau_i P_0}{\sqrt{2umkT}} \qquad (2.60)$$

τ is the molecular residence time for the first layer and τ_i is the molecular residence time for subsequent layers.

For isotherms having sharp knees, $\exp(-ax) \ll 1$, and the equation for multilayer adsorption simplifies to:

$$V = V_m \exp(bx) \tag{2.61}$$

These equations have been used to interpret experimental data with some success [84].

2.8 Hüttig equation

Several attempts have been made to extend the scope of the BET equation. Hüttig [85] assumed that the evaporation of the ith layer molecule was unimpeded by the presence of molecules in the $(i-1)$th layer whereas in the BET derivation it is assumed that they are completely effective in preventing the evaporation of underlying molecules. Hüttig's final equation is:

$$\left(\frac{V}{V_m}\right) = \frac{cx}{(1+cx)}(1+x) \tag{2.62}$$

i.e.

$$\frac{x}{V}(1+x) = \frac{1}{cV_m} + \frac{x}{V_m} \tag{2.63}$$

A plot of the left hand side of this equation against x should be a straight line of slope $(1/V_m)$ and intercept $(1/cV_m)$. Theory and experiment agree to $x = 0.7$ but at higher pressures Hüttig's equation predicts too low an amount adsorbed whereas the BET equation predicts an amount that is too high [86]. For the majority of gas–solid systems V_m calculated from Hüttig's equation exceeds the BET value by from 2% to 20% depending on the value of c. Compromise equations between Hüttig and BET have also been attempted [16].

2.9 Harkins and Jura relative method

Harkins and Jura [87] derived an equation, by analogy with condensed liquid layers, independent of V_m hence avoiding any explicit assumption of the value of molecular area of the adsorbate in the calculation of surface area.

Condensed monolayers on water are characterized by the fact that they exhibit a linear pressure-area relationship:

$$\pi = b + \alpha\sigma \tag{2.64}$$

where π and σ are the pressure and mean area per molecule, while a and b are constants.

The relationship persists up to high pressures where the film is several molecules thick. They transformed this relationship into an equivalent equation:

$$\ln(P) = B' - A'/V^2 \tag{2.65}$$

$$\ln(x) = B - A/V^2 \tag{2.66}$$

where B is a constant of integration and:

$$A = \frac{10^{20}\, aS^2 M^2}{2RTN} \tag{2.67}$$

a is a constant, S is the surface area of the solid, M is the molar gas volume, R is the molar gas constant, T the absolute temperature and N is the Avogadro's constant.

Equation (2.65) involves only the quantities P and V which are measured directly in the experimental determination of adsorption. Harkins and Jura reported that this simple equation was valid over more than twice the pressure range of any other two-constant adsorption isotherm equation.

A plot of *log x* against *1/V²* should give a straight line of slope $-A$ which is directly related to the surface area of the solid by the relationship:

$$S = k\sqrt{A} \tag{2.68}$$

where k is a constant for a given gas at a constant temperature.

This constant has to be determined by calibration, using an independent method for surface area determination. For this reason the method is known as the Harkins and Jura relative or HJr method. The original determinations were carried out with anatase, whose area had been determined from heat of wetting experiments. Orr and Dallavalle

Table 2.2 Values of HJr constants

Gas	Temperature (°C)	k
Nitrogen	−195.8	4.06
Argon	−195.8	3.56
Water vapor	25	3.83
n-Butane	0	13.6
n-Heptane	25	16.9
Pentane	20	12.7
Pentane–1	20	12.2

[7] have listed the values of k for some gases which, for convenience, are presented so that, if the volume V is in $cm^3\ g^{-1}$ at STP, the surface S is in $m^2\ g^{-1}$ (Table 2.2).

It was tacitly assumed that k was a function solely of the temperature and nature of the adsorbate and independent of the nature of the adsorbent; fundamentally the HJr method is empirical.

If the relationship between $\log x$ against $1/V^2$ is expressed as a line with two or more segments of different slopes then, according to Harkins and Jura, the slope in the low pressure region should be used, since this is the region in which the transition from monolayer to multilayer occurs. The presence of more than one linear portion is attributed to the existence of more than one condensed phase.

It is noted that the BET method yields a value for V_m from which the surface area can be calculated whereas the HJr method yields a surface area directly without generating a value for V_m.

2.10 Comparison between BET and HJr methods

Livingstone [88,89] and Emmett [90] have found that in the linear BET region, $0.05 < x < 0.35$, a linear HJr plot is only obtained for $50 < c < 250$. For $c = 2, 5$ and 10 and $x < 0.4$ there is no linear relationship, while for $c = 100$ the range of mutual validity of the two equations is limited to the region $0.01 < x < 0.13$. Smith and Bell [91] extended this enquiry to the n-layer BET equation.

Both the BET and the HJr methods are open to criticism in that they involve the arbitrary selection of constants (k and σ) which undoubtedly depend on the nature of the solid surface.

Of the two, the HJr is relatively inferior for the following reasons [16].

- The quantity $1/V^2$ is sensitive to slight errors in V.
- The range of relative pressures over which the HJr plot is linear is variable, depending on the value of c in the BET equation. For each new solid, a large number of experimental points may be needed in order to locate the linear region.
- Some systems yield HJr plots with more than one linear section.

However, the adsorption of nitrogen at $-195°C$ on the majority of solids is characterized by c values in the range 50–240, so that the surface areas obtained by the two equations are in agreement.

2.11 Frenkel-Halsey-Hill equation

Hill [92] took into account the decrease in interaction energy for molecules adsorbed in the second and subsequent layers to derive the equation:

$$\ln x = -b\left(\frac{V}{V_m}\right)^s \qquad (2.69)$$

where *b* and *s* are constants. The equation has been discussed by Parfitt and Sing [93] and an analysis of the two constants has been carried out [94].

2.12 *V–t* curve

Oulton [95] assumed that the thickness of the adsorbed layer remained constant over the whole pressure region. More accurately *t* must be related to the amount adsorbed. Provided this relationship is known, the surface area may be determined from a plot of volume adsorbed against film thickness. The plot should be a straight line through the origin and the specific surface is obtained from the slope (see section 3.5).

The film thickness for nitrogen is given by the equation:

$$t = y\left(\frac{V}{V_m}\right) \qquad (2.70)$$

where *y* is the thickness of one layer of molecules. The value of *y* will depend on the method of stacking of successive layers. For nitrogen, if a cubical stacking is assumed, $y = \sqrt{(0.162)} = 0.402$ nm. Schüll [96] and Wheeler [97] assumed a more open packing and arrived at a value of $y = 0.430$ nm. Lippens *et al.* [98] assumed hexagonal close packing for nitrogen to give:

$$y = \frac{MV_s}{N\sigma} \qquad (2.71)$$

where

M	is the molecular weight of the gas;
V_s	is the specific volume;
N	is Avogadros number;
σ	is the area occupied by one molecule.

$$y = \frac{28 \times 1.235 \times 10^{-6}}{6.023 \times 10^{23} \times 16.2 \times 10^{-20}}$$

$y = 0.354$ nm

Combining with equation (2.9) gives:

$$S_t = 1.547 \left(\frac{V}{t} \right) \tag{2.72}$$

with V in cm^3 g^{-1} (vapor at STP), t the film thickness in nm and S_t the specific surface in m^{2}g^{-1}.

This relationship cannot be used for all substance and several t–curves are available. One of the most widely used is de Boer's common t–curve. The experimental points, for a variety of substances, deviate by 10% or more from the average curve. When the t–curve is being used for pore size distribution these errors are small enough to be neglected. For accurate surface area determination, t–curves with small deviation are required, and these should be common for groups of materials such as halides, metals, graphite [99]. Lecloux [100] and Mikhail *et al.* [101] however suggest a dependency on the BET c constant rather than on the type of material (see section 3.5 for a fuller discussion).

The application of the t–curve method has been challenged by Pierce [102] and Marsh *et al.* [103,104], who state that once a molecule is adsorbed into a micropore it fills spontaneously, thus leading to the unrealistically high surface areas found in some activated carbons.

Kadlec [105] states that the t–method yields incorrect values for the specific surface of mesopores at pressures below the closure pressure of the hysteresis loop since at these pressures micropore filling has not been completed. These limitations are overcome with the t/F method where:

$$\frac{v_c}{F} = v_{micro} + (S_{me} + S_{ma}) \frac{t}{F} \tag{2.73}$$

v_c is the condensed volume adsorbed, F is the degree of volume filling, v_{micro} is the volume of micropores and S_{me} and S_{ma} are the surface areas of the meso and macropores. Kadlec also proposed a method of evaluating F which was later applied by Dubinin *et al.* [106].

2.13 Kiselev's equation

A criterion as to the correctness of the BET equation is its agreement with other models such as that derived from the Kelvin equation for pressure lowering over a concave meniscus. This has been applied to a porous adsorbent as follows:

$$RT \ln x = \frac{-2 \gamma V_m}{r_k} \tag{2.74}$$

If a_H is the number of moles adsorbed at the beginning of the hysteresis loop when the relative pressure is x_H, a small increase in pressure will lead to Δa moles of adsorbate being adsorbed where:

$$\Delta a = \frac{\Delta V}{V_m} \qquad (2.74)$$

ΔV is the liquid volume of adsorbate being adsorbed. This volume will fill the cores of radius r_c and surface area ΔS.

For cylindrical pores:

$$r_c \Delta S = 2\Delta V$$

so that:

$$\Delta S = \frac{RT}{\gamma} \ln x \Delta a$$

This may be written:

$$S = \frac{RT}{\gamma} \int_{a_H}^{a_s} \ln(x)\mathrm{d}a \qquad (2.75)$$

where a_s is the number of moles adsorbed at saturation pressure. This equation was derived by Kiselev [107].

Brunauer [108] used the following form of equation (2.75), which he claimed reduces the equation to a modelless equation as opposed to the adoption of cylindrical pores:

$$r_H \Delta S = 2\Delta V \qquad (2.76)$$

where r_H is the hydraulic radius. For the adsorption branch of the isotherm, a cylindrical pore model yields $r_H = r_K = 2r_c$ and for the desorption branch $2r_H = r_K = 2r_c$.

There is therefore no difference between Brunauer's modelless model and the cylindrical pore model uncorrected for residual film. They both yield the surface and volume of the cores of the pores which will approach the BET values for materials having only large diameter pores. Brunauer and Mikhail [109] argue that the graphical integration of equation (2.65) is more accurate than previous tabular methods.

In a more rigorous derivation Kadlec [110] replaces γ in equation (2.65) with $\gamma_{LV} + \gamma_{SV} + \Delta\pi$ where the suffixes refer to liquid–vapor and solid–vapor interfaces and $\Delta\pi$ is the difference between the spreading pressure at P_0 and at the given pressure. He further states that the beginning of hysteresis is not connected with the lower limit of applicability for mesopores [111].

2.14 Experimental techniques

2.14.1 Introduction

The available commercial equipment can be divided into three types. These may be characterized as static, volumetric units, static gravimetric units and flow, thermal conductivity units; these are all available as single point or multipoint, manual or automatic.

2.15 Volumetric methods

2.15.1 Volumetric methods for high surface areas

In all the volumetric methods the basic principles are the same. The adsorbate is degassed under vacuum to remove surface contamination. Helium is next admitted into a burette of known volume and its pressure and temperature measured so that the amount at STP can be calculated. The sample tube is immersed in liquid nitrogen and helium admitted. The residual amount in the burette is determined and the amount expanded into the sample tube determined. Since helium does not adsorb on to the solid, this volume is termed the dead space volume and it is found to be linearly dependent on pressure. The helium is removed and the procedure repeated with nitrogen. When the nitrogen expands into the sample tube, it splits into three parts, residual in the burette, dead space which can be calculated from the previously found dead space factor, and adsorbed. The process is repeated at increasing pressures and the amount adsorbed determined as a function of relative pressure.

The accurate determination of the amount of gas adsorbed depends upon a precise knowledge of the dead space. Estimation of the quantity of unadsorbed gas in the dead space is complicated by the fact that part of the dead space is at room temperature and part is at liquid nitrogen temperature.

Since the amount adsorbed represents the difference between the amount admitted to the burette and the amount in the dead space at equilibrium, it can only be evaluated with confidence if the two quantities are of unlike magnitude. To achieve this the apparatus is designed to minimize dead space volume. In practice it is found convenient to fix the volume and temperature and measure the pressure.

Regardless of the particular design, the basic apparatus must provide means for removing gases and vapors which all material pick up when exposed to the atmosphere. The apparatus must also provide means for permitting re-adsorption of known quantities of vapor on the material. It should also have evacuating systems, gauges to measure vacuum, a gas storage part and an analytical part.

A great variety of volumetric apparatus have been described in the literature and the earlier ones have been reviewed by Joy [112]. Conventional types of nitrogen adsorption apparatus follow the design described by Emmett [113].

The main disadvantage of the original design is that the sample tube is not directly connected to the vacuum line and hence any powder flying from the tube is likely to contaminate the whole apparatus. Improved designs have been described by Emmett (1944) [114], Harvey (1947) [115], Joyner (1949) [116], Schubert and Koppelman (1952) [117] , and Vance and Pattison (1954) [118].

A number of refinements have been suggested either to increase the accuracy or reduce the tedium of measurements. For example, Vance and Pattison [118] used a magic eye electrical zero point device for the manometer. Harkins and Jura [119] used a narrow bore mercury cut-off to serve as a null point instrument, the absolute pressure being measured on a wide bore manometer. Several authors have shown how the function of manometer and burette can be combined in a single device [120,121].

It has been suggested that the adsorption of mercury vapor could affect adsorption of nitrogen and to overcome this problem Dollimore *et al.* [122] devised a doser unit incorporating a pressure transducer to replace the mercury manometer. Bugge and Kerlogue [123] simplified the apparatus by using only one bulb instead of several, but with a loss of versatility. They also gave a simplified method of calculation to eliminate the dead space determination. It appears that this procedure is only satisfactory with high *c* values [112]. Several authors have used oxygen or nitrogen thermometers for the accurate measurement of the saturation vapor pressure of nitrogen [118,123]. . Lippens, Linsen and de Boer [124] state that none of the above apparatus is suitable for accurate determination of pore size distribution determination and describe an apparatus which fulfills the basic criteria:

- rapid removal of heats of adsorption and supply of heats of adsorption to give rapid equilibrium;
- clear establishment of equilibrium;
- recovery of the total amount of nitrogen adsorbed in order to check for leakage;
- temperature changes of the liquid nitrogen bath monitored continuously;
- facility to suspend measurements in order to eliminate the need for overnight supervision.

Recently an automatic apparatus for surface areas and pore size determination has been described [125]. The Isothermegraphe is a volumetric apparatus, with a calibrated tube, which draws complete adsorption–desorption isotherms using a piston of mercury which modifies the pressure slowly at a programmed speed.

Jaycock [126] determined the optimum volumes for the gas burettes to give six evenly spaced points on the BET plot and this design was incorporated in the British Standard [127].

2.15.2 Volumetric methods for low surface areas

Surface areas down to 1 $m^2 g^{-1}$ can be determined using nitrogen adsorption provided great care is taken. A semi-micro unit has been described for surface area determination down to 2 $m g^{-1}$ [128] and a capillary differential manometer to keep the dead space low [129]. Since the amount of gas in the dead space is proportional to the absolute pressure, it is preferable to use gases with low saturation vapor pressures. The adsorption equipment is very similar to that used for measurements with nitrogen, the only difference being the pressure range of the gauge. Several types of apparatus have been described [130-133].

Wooton and Brown [134] used this low pressure method to measure the surface area of oxide coated cathodes (*c*. 100 cm^2) by adsorption of ethylene and butane at −183°C and −116°C respectively. Because of the very low pressures involved in the technique no leaks could be tolerated. The apparatus was made, therefore, entirely of glass, mercury cut-offs were used instead of stopcocks and the sample chamber was welded on to the system to eliminate any possibility of leaks. A dry ice trap between the sample and the mercury cut-offs served to prevent mercury vapor from reaching the sample. Equilibrium pressures were measured with a highly sensitive McLeod gauge.

Lister and McDonald [135] have described in detail the construction and calibration of low temperature ethylene adsorption apparatus. In measurements of such low pressures, two obvious risks must be considered namely the desorption of water and thermal transpiration. By heating the entire system for a short while, or by permanently keeping the system under vacuum, most of the adsorbed vapors from the glass walls should be removed. Otherwise the slowly desorbing vapors will increase the pressure in the system during adsorption measurements leading to erroneous results. When low pressure measurements are made on a gauge held at a different temperature from that part of the apparatus where the adsorption takes place, correction for thermal molecular flow also needs to be considered. To obtain accurate results Lister and McDonald prepared and used correction data.

In most low pressure measurements the correction for unadsorbed gas is negligible so that no effort needs to be made to minimize the dead space volume.

Aylmore and Jepson [136] used a novel method of krypton adsorption with labeled krypton (^{85}Kr) as adsorbate and, from the measure of activity, they calculated pressures.

The field of low surface area determination has been reviewed by Choudhary [137].

2.16 Gravimetric methods

Gravimetric methods have the great advantage over volumetric ones in that the volume of the adsorption system is immaterial and the amount of gas adsorbed is observed directly by measuring the increase in the weight of the solid sample upon exposure to a gas or a vapor. The tedious volume calibration and dead-space determinations are thus eliminated.

The main disadvantages of the method are:

- the apparatus is much less robust and correspondingly more difficult to construct and maintain than volumetric apparatus;
- the apparatus has to be calibrated by placing known weights in the adsorbent pan, and the method is hence subject to the errors always attached to determinations which are dependent on the constancy of calibrations of easily fatigued and strained mechanical systems;
- buoyancy corrections have to be made.

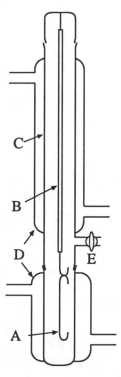

Fig. 2.3 The McBain Bakr spring balance.

Thus, although many different types of gravimetric apparatus have been reported, they have not become popular due to their delicate nature and the difficulty of compensating for buoyancy effects.

In gravimetric methods of surface area determination the amount of gas adsorbed is determined by weighing using an adsorption balance. A sensitivity of around 1 μg with a load of 5 g is required. Beam micro-balances are commercially available based on a design by Cahn and Schultz [138]. Helical spring balances based on a design by McBain and Bakr [139] have also been widely used. More recently balances based on magnetic and linear differential transformers have become available.

2.16.1 Single spring balances

McBain and Bakr [139] introduced a sorption balance (Figure 2.3). This consists of a glass vessel in three parts C which is maintained at constant temperature at D. The essential features of the balance are a quartz helical spring B supporting a small gold or platinum bucket A. The spring is calibrated by adding small known weights to the bucket and measuring the increase in length of the spring with a catherometer. The bucket is filled with the adsorbent and the entire apparatus outgassed through valve E. Adsorbent is then introduced and the extension of the spring measured. Buoyancy corrections need to be applied since the gas density changes with pressure.

This type of balance is restricted in use to condensable adsorbates and is especially useful at higher pressures. Morris and Maass [140], Dunn and Pomeroy [141] and McBain *et al.* [142,143] have used similar apparatus. Several others have used single spring balances with improvements and modifications to suit their applications. Boyd and Livingstone [144] used mercury cut-offs in the vapor handling and compressing systems. The pressure was controlled by compressing the gas in the dosing bulb and reading the pressure on a mercury manometer or McLeod gauge, depending on the range. Seborg, Simmons and Baird [145] dried the sample in a current of dry air, and obtained the adsorption points subsequently by passing the air through saturators filled with solutions of known vapor pressure. Dubinin and Timofeev [146] used a magnetically operated greaseless doser for the admission of precise amounts of adsorbate increments. Automatic recording techniques have also been described [147].

2.16.2 Multiple spring balances

Gravimetric methods have an advantage over volumetric ones in that several determinations can be carried out simultaneously by connecting several balance cases to the same gas or vapor manifold and observing the individual spring extensions.

Seborg and Stamm [148] connected five or six simple spring units in series. Pidgeon and Maass [149]. Mulligan *et al.* [150] and Stamm and Woodruff [151] connected as many as sixteen springs to the same apparatus.

2.16.3 Beam balances

Beam vacuum microbalances have greater sensitivity than helical spring balances and the troublesome buoyancy correction at high pressure is eliminated, at least partly if not completely.

Beam balances can be of either high sensitivity at very low total loads or of medium sensitivity at large total loads, which is in contrast to the normal short spring balances which have a medium sensitivity at low total loads.

The majority of the high sensitivity low-load balances are based on those originally designed by Barrett, Birnie and Cohen [152] and by Gulbransen [153]. Barrett *et al.* used a glass beam 40 cm long supported on a tungsten wire and enclosed the whole assembly in a tubular glass casing connected to the vapor and vacuum manifolds. Calibration was effected by moving a small soft-iron rider along the beam by means of a magnet outside the case. Gulbransen's balance was constructed from glass rod, quartz wires and metal wires on the same principles as an ordinary chemical balance.

Rhodin's microbalance [154–156] is essentially a modification of these, in which some stability is sacrificed for increased sensitivity, by the use of thinner and lighter wires. This balance was adopted by Bowers and Long [157] for adsorption at liquid helium temperatures.

Rhodin's balance was made as symmetrical as possible in order to eliminate buoyancy corrections and to minimize thermal eddy currents. The adsorbent and counter weights were matched to within 10^{-5} g and immersed to the same depths in identical thermostatic baths and the outgassing was done at 400°C in a vacuum of 10^{-7} mm Hg. With this

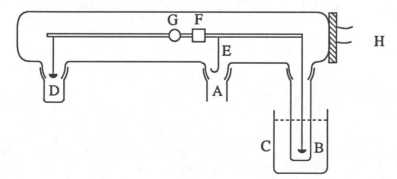

Fig. 2.4 Line diagram of the Cahn vacuum microbalance.

balance it was possible to observe a vertical displacement of 10mm to better than 0.01 mm and with loads of up to 1 g it was possible to observe weight changes of 10^{-7} g in a reproducible manner.

Beam balances have also been operated as null-point instruments. The beam is acted upon by a solenoid outside the balance housing, the current through the solenoid being adjusted to restore the beam to its horizontal position. One such balance by Gregg [158] used two concentric solenoids, the inner one suspended from the beam and the outer one fixed to the envelope. The original balance had a sensitivity of 0.3 mg, the load being as high as 10 to 20 mg. In an automatic version of this instrument, described by Gregg and Wintle [159], a photoelectrically operated relay adjusted a potentiometer slide wire contact which was connected to the solenoid on the balance.

The Cahn [160] microbalance illustrated in Figure 2.4 has a sensitivity of 1 µg with a load of 0.5 g. The system is evacuated through manifold A. The adsorbate is contained in bucket B which is immersed in a nitrogen bath C. Counterbalance is provided at D with provision for larger weights through hook E. Movement of the beam is sensed by photocell F which provides feedback to coil G to restore the beam to its equilibrium position. The weight changes are transmitted to a recording device through electrical connections at H.

2.17 Flow, thermoconductivity methods

Lobenstein and Dietz developed an apparatus not requiring a vacuum system [161]. They adsorbed nitrogen from a mixture of nitrogen and helium in two burettes by continuously raising and lowering attached mercury columns. Equilibrium was established when constant pressure was attained. Additional points were obtained by adding more nitrogen to the system.

The method, a modification of gas adsorption chromatography in which the column packing is the sample itself and the mobile gas phase is a mixture of a suitable adsorbate and an inert gas, was further developed by Nelson and Eggertsen [162]. They used nitrogen as the adsorbate and helium as the carrier gas. They also used cathetometers to improve accuracy.

A known mixture of helium and nitrogen is passed through the sample and then through a thermal conductivity cell connected to a recording potentiometer. When the sample is cooled in liquid nitrogen it adsorbs nitrogen from the mobile phase; this is indicated by a peak on the recorder chart, and after equilibrium is attained the recorder pen resumes its original position. Removing the coolant gives a desorption peak equal in area and in the opposite direction to the adsorption peak. Either peak may be used to measure the amount of nitrogen adsorbed but the latter tends to be better defined.

Calibration for such a system may be absolute (by injecting a known amount of adsorbate into the mobile phase at the point normally

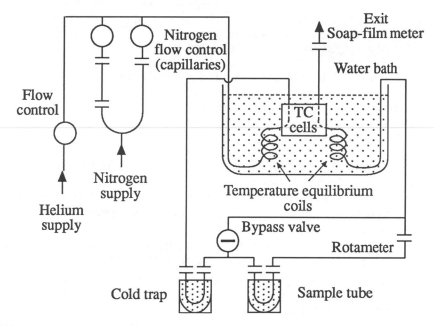

Fig. 2.5 Schematic diagram of Nelson and Eggertsen's apparatus.

occupied by the sample and noting the area under the peak) or by comparison with a sample of known area.

A schematic diagram of their apparatus is shown in Figure 2.5. Nitrogen flow control was achieved by the use of two capillary tubes in parallel, 0.25 mm internal diameter and 150 and 300 mm long. The capillaries were used independently or together to give three nitrogen flowrates in the range 5-20 mL min^{-2} with a pressure head of 2 lb in^{-2}. The helium flowrate was controlled by needle valves. The flowrates were measured by rotameter and soap-film meter.

The mobile phase was first passed through the reference arm of the thermal conductivity cell on to the sample and then to the measurement arm of the thermal conductivity cell.

Nelson and Eggertsen measured adsorption at three concentrations i.e. three partial pressures. The sample was outgassed at the desired temperature (up to 500°C) for 1 h while being purged with helium flowing at 20 mL min^{-1}. Nitrogen relative pressures in the range 0.05 to 0.35 and a total nitrogen flowrate of 50 mL min^{-1} were used. The desorption peaks were used for measurement purposes because they were relatively free from tailing effects.

Calculation is essentially the same as for the pressure–volume method but is much simpler since no dead space correction is required. The authors assumed complete linearity of the thermal conductivity cell over the concentration range employed. They analyzed samples ranging in surface area from 3 to 450 $m^2 g^{-1}$ and the results were in good agreement with static volume data.

The advantages over the conventional BET method are:

- elimination of fragile and complicated glassware;
- elimination of a high vacuum system;
- permanent records obtained automatically;
- speed and simplicity;
- elimination of dead space correction.

Ellis, Forrest and Howe [163] made modifications and improvements to the original design for their specific application. All flow controls were done by needle valves and all flow measurements by rotameters. They used a helium flowrate of 50 mL min^{-1} and nitrogen flowrates of 3, 5 and 10 mL min^{-1}.

By taking more care when chilling the sample, to avoid shock effects in the gas stream, they extended the measurement range down to 0.01 $m^2 g^{-1}$ and obtained good linearity in the BET plot, even at such low surface areas. Their published data were on samples in the specific surface range of 0.005 to 14.2 $m^2 g^{-1}$.

Below 0.01 $m^2 g^{-1}$ conventional adsorption methods were considered unreliable and no comparisons were given by them. Above that range agreement between continuous flow data and volumetric data were good with the former data slightly higher.

Ellis *et al.* developed a shortened method using only one flow rate, i.e., a single point method. They analyzed a number of samples and determined the adsorption peak area for 10 mL min^{-1} flow of nitrogen and 50 mL min^{-1} helium obtaining a linear graphical relationship from which the surface areas of subsequent samples were obtained. Results were again comparable with conventional volumetric BET measurements.

Atkins [164] developed a precision instrument for use in the carbon black industry. For precision measurements he stated that it was necessary to consider changes in ambient temperature, barometric pressure, liquid nitrogen temperature and nitrogen contamination in the gas mixture. Correction for non-linearity of the catherometer is also necessary. Atkins used heat exchanger coils in the detector circuit in addition to temperature control of the detector.

Haley [165] extended the continuous flow measurements to include the size distribution of pores in the 1 to 30 nm radius range using 10% nitrogen in helium at various pressures up to 150 lb in^2, causing the nitrogen to reach its liquefaction point. The amount of nitrogen adsorbed or desorbed was measured continuously. He also measured surface areas obtaining a variation of ±2.5% in the range 40 to 1250 $m^2 g^{-1}$.

Since helium is expensive it may be replaced by other gases which are not adsorbed under experimental conditions, e.g. hydrogen.

Perkin-Elmer Ltd developed a commercial continuous flow apparatus, the Sorptometer. The manufacturers claimed that a three point determination of surface areas in the 0.1 to 1000 $m^2 g^{-1}$ could be carried out in 20 to 30 min using pre-mixed gases. Degassing was carried out using a gas purge [166].

Atkins [163], using this apparatus, obtained a relative standard deviation varying from 1.76% to 2.99% according to sample material, with ten single point determinations, each with a new sample. With his own equipment the standard deviation varied between 0.25% and 1.35%.

Several problems arose when the Nelson and Eggertsen type apparatus were used for surface area determination smaller than 500 $cm^2 g^{-1}$ due to distortion of the adsorption and desorption peaks.

During adsorption an adsorption reverse peak is produced when the sample tube is immersed in liquid nitrogen. This immersion causes a contraction of the gas inside the sample tube and a reduction in the gas flowrate through the catherometer. The thermister on the measurement side warms up causing a change in its resistance and a peak on the recorder chart. A reverse peak occurs midway through the adsorption.

During the desorption cycle a desorption reverse peak occurs immediately prior to desorption. This is followed by a peak of the same area but different sign midway through desorption. Since the desorption peak is used for measurement purposes, effort has been directed mainly at finding an explanation for the desorption reverse peak and correcting for it.

Lowell [167 considered that the reverse peak on desorption was due to transverse thermal diffusion. When the sample tube is removed from the coolant, pre-cooling of the gas stream in the inlet section quickly ceases and gas enters the sample catching section at very nearly room temperature. Since the sample is still cold, partial separation of the gases takes place due to transverse thermal diffusion, nitrogen moving to the walls of the container and helium to the center. Gas flow is more rapid at the center and helium rich gas is carried to the catherometer, giving rise to a thermister cooling peak. Immediately afterwards, when the temperature of the sample tube has risen considerably, the separated nitrogen diffuses into the gas stream and produces a peak in the opposite direction and necessarily of the same area. Thus, the area of the desorption peak is increased by the area of the reverse peak. The true area of the desorption trace is the area of the observed peak minus the area of the reverse peak.

Lowell's experimental solution was to allow the desorbed gases to expand into a vessel whose volume could be adjusted by altering the amount of mercury in it. The container consisted of two cylinders containing mercury, one closed and one open, connected at the bottom by a flexible tube. As the gas expands into the closed tube it forces the mercury down and the pressure is equalized by lowering the second container. When the desorption and adsorption processes are complete and enough time has elapsed for mixing to take place, the gas is swept through the thermal conductivity cell and the area of the desorption peak is measured with no interference from a reverse peak.

Tucker [168] removed the anomalous peaks by using an interrupted flow technique. A simple [169] and a modified [170] continuous flow apparatus have also been described.

2.18 Sample preparation

2.18.1 Degassing

A most important preliminary to the accurate measurement of an adsorption isotherm is the preparation of the adsorbent surface. In their usual state all surfaces are covered with a physically adsorbed film which must be removed or degassed before any quantitative measurements can be made. As the binding energy in physical adsorption is weak van der Waals forces, this film can be readily removed if the solid is maintained at a high temperature while under vacuum.

The degree of degassing attained is dependent on three variables, pressure, temperature and time. In test and control work, the degassing conditions may be chosen empirically and maintained identical in all estimates since only reproducibility is required. For more accurate measurements, conditions have to be chosen more carefully.

2.18.2 Pressure

Although it is advisable to outgas at as low a pressure as possible, due to considerations of time and equipment the degassing pressure is kept as high as is consistent with accurate results. The pressures usually recommended are easily attainable with a diffusion pump. Emmett [171], for example, recommends 10^{-5} mm Hg, while Joy [112] recommends 10^{-4} mm Hg, since under this condition the rate of degassing is controlled largely by diffusion from the interior of the particles.

For routine analyses Bugge and Kerlogue [172] found that a vacuum of 10^{-2} to 10^{-3} mm Hg was sufficient and the differences in surface areas so obtained was smaller than 3% of those obtained at 10^{-5} mm Hg.

Vacuum should be applied slowly to prevent powder from being sucked into the vacuum line. Cleaning contaminated equipment is a time-consuming chore and contamination can be prevented provided care is taken. The introduction of a plug of cotton wool into the neck of the sample tube can reduce the possibility of powder loss. At the end of the degassing cycle the sample cell is isolated from the vacuum for about 15 min; degassing is deemed complete if no pressure increase can be detected when the cell is reintroduced to vacuum.

Since vacuum is not applied with continuous flow apparatus, degassing is effected by purging the sample using the carrier gas as the purge gas. The gas needs to be of high purity and is passed through a cold trap to reduce traces of organic compounds or water vapor. Completion of degassing is determined by passing the effluent over a thermal conductivity cell.

2.18.3 Temperature and time.

Recommended temperatures and times for degassing vary considerably in the literature and it is difficult to establish any single degassing condition acceptable for all solids. However, Orr and Dallevalle [7] give an empirical relationship which they suggest is acceptable as a safe limit for ordinary degassing at pressures lower than 5×10^{-6} mm Hg.

$$\theta = 14.4 \times 10^4 t^{-1.77} \tag{2.77}$$

where θ is in hours and t in °C (applicable between 100°C and 400°C). This can only be taken as a general safe limit since many others have found the necessary time much shorter.

Holmes *et al.* [173] determined the surface area of zirconium oxide using argon, nitrogen and water vapor as adsorbates. They found that the surface properties depended upon the amount of irreversibly adsorbed water which was far in excess of a monolayer. Degassing at 500°C resulted in a 20% decrease in area.

McBain and Bakr [139] recommends an adsorption–desorption cycle to reduce the time of degassing. He flushes the sample with adsorbate at the temperature of the forthcoming measurements, followed by heating it under vacuum.

For samples which degrade at elevated temperatures, repetitive adsorption–desorption can also be employed to clean up the surface. Usually three to six cycles are required to obtain reproducible data [174].

2.18.4 Adsorbate.

Nitrogen and argon are the most commonly used adsorbates but any non-corrosive gas can be used without special calibration: He, H_2, O_2, CO_2, CO, etc., or Kr using a special krypton unit.

The standard technique is the adsorption of nitrogen at liquid nitrogen temperature, evaluation being by the BET equation in the approximate relative pressure range $0.05 < x < 0.35$.

Using the volumetric method, for powders of reasonably high surface areas (greater than 10 m^2 g^{-1}), the proportion of the admitted gas which is adsorbed is high. The consequent change in pressure, due to adsorption and expansion into the 'dead space' can be measured accurately with a mercury manometer.

For powders of low surface area (< 5 m^2 g^{-1}) the proportion adsorbed is low; most of the gas introduced into the sample tube remains unadsorbed in the 'dead space' leading to considerable error in the determined surface area. The use of krypton or xenon at liquid nitrogen temperature is preferred in such cases since the low vapor pressure exerted by these gases greatly reduces the 'dead space' correction factor thus reducing the error. In addition, the pressures encountered are low enough that the deviations from perfect gas relations are negligible.

Evaluation of surface area using these gases is complicated since the area occupied by the molecule varies with the adsorbent $[0.17 < \sigma_{Kr} < 0.23]$ nm^2, $[0.17 < \sigma_{Xe} < 0.27]$ nm^2.

Beebe *et al.* [175] recommend the use of krypton at liquid nitrogen temperature which, due to its low saturation vapor pressure, reduces the amount of unadsorbed gas in the gas phase. Beebe's value of 0.185 nm^2, for the area occupied by a krypton molecule is preferred by most investigators [176–178] but 0.195 nm^2, has also been quoted [179] There is also disagreement over the correct saturation vapor pressure to use. The use of the solid saturation vapor pressure of 1.76 torr at 77.5 K usually results in the production of markedly curved plots [180]. Later investigators [181] tended to use the extrapolated vapor pressure of 2.63 torr.

[Note: To convert pressure p in mm Hg (torr) at temperature T to Pascals, it is first necessary to convert to pressure p' at STP by correcting for thermal expansion $p' = p[1 - 1.82 \times 10^{-4}(T - 273.15)$ and multiply the results by the conversion factor 133.322. Thus 64 mm Hg pressure at 25°C = 8494 Pa].

The traditional McLeod gauge used with nitrogen adsorption is replaced by a thermocouple or thermister in order to measure these low pressures.

The saturation vapor pressure of argon at liquid nitrogen temperature is 187 torr which is lower than that for nitrogen, hence the amount of unadsorbed vapor is reduced leading to a better estimate of monolayer volume. However the area occupied by argon molecules is governed by adsorbate–adsorbent interactions, hence it varies from system to system.

For the adsorption of argon on boron carbide, Knowles and Moffat [182] found that the application of the BET theory gave more consistent results using liquid argon pressure rather than the solid-vapor pressure.

Young and Crowell [16] have listed the molecular areas of many adsorbates. In practice, for consistency, the areas are corrected on the basis of the area occupied by a nitrogen molecule at liquid nitrogen temperature. However, the area occupied by a molecule may depend upon the nature of the surface and calibration for that particular solid-vapor system may be necessary [183].

Nitrogen adsorption is governed by adsorbate-adsorbate interactions, particularly near the completion of a monolayer. This lateral interaction pulls the molecules together to form a close-packed liquid like monolayer. An exception [184] is found with graphitized carbon on which the nitrogen molecule occupies 0.200 nm^2, rather than the more usual 0.162 nm^2 i.e. one nitrogen molecule to three carbon hexagons. The lateral interaction, in this case, is not strong enough to pull the nitrogen molecules together. This can only happen on high energy level surfaces. The specificity of nitrogen and water vapor on hydroxylated and dehydroxylated silicas has also been investigated. Non-, meso- and micro-porous silicas were examined using BET, FHH and α_s methods [185].

In a review of gas adsorption literature Avnir [186] found that, in every case, the area measured using different size molecules increased with decreasing molecule size. This is in accord with the concept of fractals; a plot of measured surface against molecular size on a logarithmic scale yielding a straight line, the slope of which is a measure of surface roughness.

Ethylene (SVP = 0.1 mm Hg) at liquid oxygen temperature has also been used for low surface area determination [187,188].

2.18.5 Interlaboratory tests

The goal of analysis is high precision within a laboratory and high reproducibility between laboratories. Desbiens and Zwicker [189] carried out interlaboratory tests with alumina and found that the degassing temperatures and times were critical. AFNOR [190] also carried out inter-laboratory tests with alumina and found wide disparities between laboratories. It cannot be stressed too strongly that commercial equipment should be calibrated against standard equipment at regular intervals. An alternative is to calibrate with known standards and control chart the results for early detection of drift.

2.19 Standard volumetric gas adsorption apparatus

A full description of this apparatus and its operating procedure is given in British Standard BS 4359 Part 1, 1984. In the apparatus illustrated (Figure 2.6), the main vacuum line consists of a 15 mm bore glass tube to which are attached the adsorption unit, a vacuum gauge and vacuum pumps. A 4L flask containing nitrogen and a 1L flask containing helium are connected to a secondary line which is joined to a gas burette, a sample tube and a mercury manometer. The gas burette consists of five carefully calibrated bulbs, with volumes (in milliliters) approximately equal to the values shown in Figure 2.7, enclosed in a water jacket. The volume of the gas contained in the burette can be adjusted by raising the level of mercury to any one of five fiduciary marks and the pressure can be read off the mercury manometer. The sample tube is connected to the gas burette through a ground glass joint. The sample tube (Figure 2.8), usually of about 2 mL volume, is specially designed to prevent 'spitting' during degassing. A third vacuum line controls the vacuum in the mercury reservoir of the gas burettes, which may also be opened to the atmosphere in order to raise or lower the level of the mercury. The entire system is evacuated by a mercury diffusion pump backed by a rotary pump capable of an ultimate vacuum of 10^{-6} mm Hg. A small electrical furnace is used to heat the sample tube while the pump is degassed. The equipment is designed to operate at maximum accuracy with 10 m^2 of powder in the sample tube, but with careful calibration surface areas as low as 1 m^2 can be determined with an accuracy of better than 10%.

In all volumetric methods, the principle underlying the determination is the same. The powder is heated under vacuum to drive off any adsorbed vapors. The pressure, volume and temperature of a quantity of adsorbate [nitrogen gas usually although krypton is used for low surface areas ($S_w < 1$ m^2 g^{-1})], are measured and the amount of gas present is determined. Traditionally this is recorded as cm^3 at standard temperature and pressure (STP) although some prefer moles.

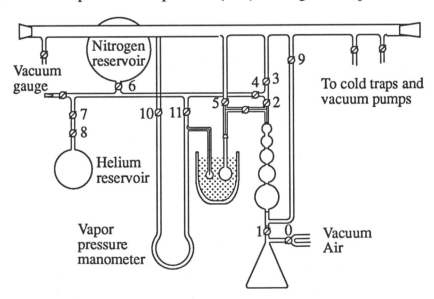

Fig. 2.6 BS 4359 Standard gas adsorption apparatus.

The adsorbent (powder) is then brought into contact with the adsorbate and, when constant pressure, volume and temperature conditions show that the system has attained equilibrium, the amount of gas is again calculated. The difference between the amount of gas present initially and finally represents the adsorbate lost from the gas phase to the adsorbate phase. The accurate determination of the amount of gas unadsorbed at equilibrium depends upon the accurate determination of the dead space or the space surrounding the adsorbent particles. The dead space is determined by expansion measurements using helium, whose adsorption can be assumed to be negligible.

Estimation of the quantity of unadsorbed gas is complicated by the fact that part of the dead space is at room temperature and part at the temperature of the adsorbate.

Since the amount adsorbed represents the difference between the amount admitted to the dead space and the amount remaining at equilibrium, it can only be evaluated with confidence when the quantities are of unlike magnitude. To achieve this, the apparatus is designed so as to minimize the dead space volume.

2.19.1 Worked example using BS 4359 standard apparatus

The volumes of the burettes (in mL) are predetermined by filling with mercury and weighing prior to assembly. This needs to be repeated several times for the necessary accuracy.

$$V_1 = 13.0, V_2 = 30.4, V_3 = 53.2, V_4 = 103.3, V_5 = 250.2$$

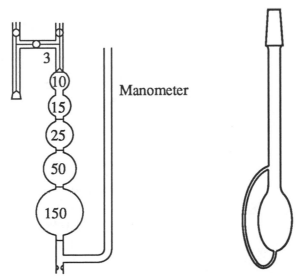

Manometer

Fig. 2.7 Gas burettes for the static BET method. Numbers correspond to bulb volumes in mL.

Fig. 2.8 Sample tube for static BET method.

(a) Calculation of dead space factor

The sample is weighed in the pre-tared sample tube which is then attached to the apparatus, and the system is then evacuated. The heating furnace is placed around the sample tube and degassing proceeds under vacuum. The degassing conditions should be carefully selected; in many cases degassing at 150°C for 1h is sufficient but degassing time is reduced if a higher degassing temperature is permissible.

Degassing is deemed complete when no discernible vapor is given off by the powder; this is detected on the vacuum gauge by closing taps 4 and 5 and isolating the sample for 15 min. If the vacuum gauge records no pressure rise when tap 5 is opened the furnace can be removed.

The furnace is then replaced by a Dewer flask containing liquid nitrogen, the level of which is kept constant throughout the analysis.

Helium is admitted to a burette (4 in this case; Table 2.3) and the gas pressure noted: knowing the pressure, volume and temperature (P,V,T), the volume admitted (V_0), at standard temperature and pressure (STP), can be calculated from:

$$\frac{PV}{T} = \frac{P_0 V_0}{T_0} \tag{2.78}$$

The mercury level is then raised to compress the volume to burette 3, and the mercury level is again noted. A third reading can also be taken at volume 2. The measured volumes should agree to 0.01 mL.

Tap 4 is opened allowing the gas to expand into the sample tube, which is immersed in the liquid nitrogen bath to a fixed, constant level, and the pressure noted at each of the burette volumes. The data are recorded in Table 2.3. The dead space factor (Q) is:

TABLE 2.3 Calculation of dead space factor.

Burette	Initial pressure	Volume of He admitted	Volume of He in burette after expansion	Final pressure
	(mm Hg)	(mL @ STP)	(mL @ STP)	(mm Hg)
4	113.2	14.096	11.307	90.8
3	219.8	14.096	9.517	148.6
2	384.8	14.102	7.659	209.0
Mean value		14.098		

$P_0 = 760$ mm Hg: $T_0 = 273.2$K: $T = 298.2$ K (*P_0 is determined prior to the analysis)

$$Q = \frac{\text{volume of helium admitted--volume in burette after expansion}}{\text{final pressure}}$$

The calculated values from Table 2.3 (mL mmHg^{-1}) are 0.0307, 0.0309, 0.0308. These should be reproducible to 0.0002.

(b) Surface area determination

Nitrogen is admitted to a burette (5 in this case) and the mercury manometer pressure reading is taken. The gas is compressed to the next burette volume and the new reading taken. As with the helium the volume is reduced to the volume at STP. Usually 2 or 3 readings are noted but more may be taken if desired (Table 2.3). Volumes should be reproducible to 0.02 mL.

The mercury is lowered to below the lowest calibration mark on the burettes and tap 4 is opened to allow the gas to expand into the sample bulb. On no account must the mercury pass this fiduciary mark prior to a reading since subsequent lowering of the pressure may not desorb the gas adsorbed at the higher pressure concomitant with this occurrence.

Table 2.4 (a). Determination of adsorbed volume of nitrogen
(b). surface area determination by gas adsorption

(a)

Burette	Initial pressure (mm Hg)	Nitrogen admitted (mL)	Final pressure (mm Hg)	(V_B)	(P_2Q)	(V_A)
				(mL @ STP)		
5	59.4	17.916	38.0	11.461	1.170	5.28
4	144.0	17.932	76.8	9.564	2.365	6.00
3	279.6	17.931	119.2	7.645	3.671	6.61
2	489.8	17.947	159.6	5.849	4.916	7.17
Mean value		17.932	214.2	3.357	6.579	7.98
		s.d. = 0.013				

(b)

Pressure (P) (mm Hg)	Volume adsorbed $V_m = V/m$ (mL g^{-1})	Relative pressure (x)	$y = \dfrac{x}{(1-x)V_m}$
38.0	2.07	0.050	0.025
76.8	2.35	0.101	0.048
119.2	2.59	0.157	0.072
159.6	2.81	0.210	0.095
214.2	3.13	0.282	0.125

weight of powder in sample tube, $m = 2.549$ g

The volume adsorbed (at STP) is calculated from:

Volume adsorbed = volume − volume in burette − dead space
admitted after expansion volume

The regression line for Figure 2.9b, generated from the data in Tables 2.4, is: $y = 0.431x + 0.00409$; Slope + intercept = $1/V_m$ hence V_m = 2.298 mL g^{-1}. Intercept = $1/cV_m$ hence $c = 106$ and:

$S_w = (10.0 \pm 0.03)$ m^2 g^{-1}

The volume adsorbed should be accurate to 0.04 mL making the measured surface accurate to 0.2 m^2.

2.20 Haynes apparatus

Hayne [191] used a vertical U-tube mercury manometer with arms of length about 60 cm connected to a sample bulb. When the bulb, containing air at atmospheric pressure, is immersed in liquid oxygen at -183°C the pressure falls by an amount ΔP_0. If the bulb contains an adsorbent solid it falls an amount ΔP, which is greater than ΔP_0 due to adsorption by the solid. A plot of $(\Delta P - \Delta P_0)/w$, where w is the weight of the adsorbent, produces a straight line when plotted against the BET surface area. The instrument is precalibrated using conventional equipment. This simple procedure is useful for monitoring a production line for a single powder.

2.21 Commercial equipment

Single and multiple point instruments are available that operate in static volumetric, continuous flow and gravimetric modes. A brief description of some of these is given below and a listing of commercial gas adsorption instruments is given in Table 2.5.

2.21.1 Static volumetric apparatus.

Essentially the static, volumetric gas adsorption equipment available commercially is for determining the amount of gas physically or chemically adsorbed on a powder surface. It is available for either single point or multipoint techniques and may be manual or automatic. Surface areas down to 1 m^2 can be determined to ±0.1 m^2 using nitrogen adsorption provided care is taken. With coarser powders the dead space errors makes nitrogen unsuitable. Since the amount of gas in the dead space is proportional to the absolute pressure it is preferable to use gases with low saturation vapor pressures. Krypton with a saturated vapor pressure of 1.76 mm Hg at −195°C is widely used.

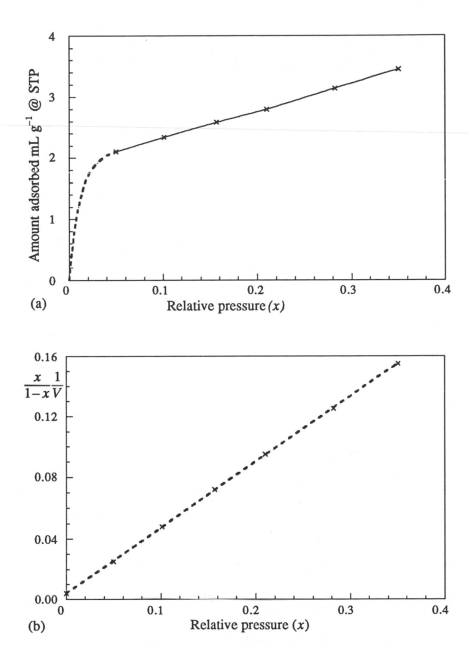

Fig. 2.9 (a) Part of adsorption isotherm showing experimental points (b) BET plot corresponding to adsorption isotherm of (a).

(a) Bel Belsorp 28

Bel manufacture the Belsorp 28 for high precision, fully automated specific surface and pore size determination. This instrument has been used with a wide variety of adsorbates including hexane and carbon dioxide [192]. The Belsorp 18 is designed for measuring the adsorption of water, organic vapors and gases.

(b) Carlo Erba Sorpty 1750

The Carlo Erba Sorpty 1750 is a typical static, volumetric apparatus, the volume of adsorbed gas being calculated by measuring the pressure change resulting from adsorption of a known volume of gas by the powder sample. The adsorbate is introduced into a variable volume chamber until it reaches a preset pressure. The chamber is then connected to a burette under vacuum containing the previously degassed sample. When the gas comes into contact with the adsorbent the gas molecules distribute themselves between the gas phase and the adsorbed phase until equilibrium is reached. From the final pressure the amount of adsorbed gas is calculated. Up to 14 routine surface area analyses can be carried out in a day.

(c) Chandler Ni-Count

This instrument is a single point analyzer designed specifically for determining the surface area of carbon black.

(d) Coulter Omnisorp series

The adsorbate flowrate into the Omnisorp sample holder is very low, therefore the first data point is obtainable at a partial pressure three to four orders of magnitude lower than that using conventional instruments. The extended data range allows critical insight into micropore characterization down to radii of 30 nm. Up to 2,000 data points per isotherm are generated: this high resolution allows pore size distribution peaks separated by as little as 20 nm to be resolved. The Omnisorp 100 is a single port instrument whereas the Omnisorp 360 analyzes one to four samples sequentially and is available with both physisorption and chemisorption capability. The Omnisorp 610 is a ten port, high volume throughput analyzer which can use a wide range of gases. A high vacuum pump pulling 10^{-6} torr, suitable for micropore analyzes, is also available.

(e) Fisons Sorptomatic 1900 Series

Having selected the sample adsorption rate and number of experimental points, the analysis is automated using an automatic gas introduction system (AGIS). AGIS can operate in up to three preselected adsorption pressure regions to give high resolution. Up to four Sorptomatics can be connected simultaneously to one computer via a multiplug RS232 connector.   The Sorptomatic has two outgassing stations with completely programmable ovens operating at temperatures up to 450°C. Multiple gas and vapor usage allows the instrument to be used in either physisorption or chemisorption mode. Specific surfaces down to 0.2 $m^2 g^{-1}$ can be measured using nitrogen and this is extended to 0.005 $m^2 g^{-1}$ with krypton.   Pore volumes are resolvable down to 0.0001c $m^3 g^{-1}$.

(f) Horiba SA-6200 Series

These use nitrogen adsorption in the continuous flow mode to measure from 0.10 to 2000 $m^2 g^{-1}$ at throughputs of up to eight analyses per hour. Model SA-6201 is a single station analyzer, SA-6202 and SA-6203 are double and triple station analyzers and SA-6210 is a sample preparation station for up to three samples simultaneously.

(g) Micromeretics accelerated surface area and porosimetry ASAP

The ASAP 2000 is a fully automated system using a wide range of adsorbate gases. This system performs single point and multi point surface area analyses, pore size and pore volume distributions completely unattended.   Specific surfaces down to 5 $cm^2 g^{-1}$ are measureable using krypton as adsorbate and pore volumes are detectable down to less than 0.0001 $cm^3 g^{-1}$. Pore size distributions are calculated using the BJH method and for micropores the Halsey or Harkins and Jura t–plots can be used.   The micropore system incorporates several micropore techniques such as the Horvath–Kawazoe, Dubunin, MP, and t–plot. Combined with these techniques is the ability to use adsorbates such as argon at liquid argon temperature as well as CO_2, N_2 and other gases.

When fitted with the krypton option the ASAP can reach a vacuum level of less than 0.00001 mm Hg which leads to accurate micropore and low surface area measurements.

The ASAP 2400 is a low cost, fully automatic version which can run six independent samples simultaneously.

(h) Micromeretics Accusorb

The Accusorb 2100E is a versatile manual unit for the determination of adsorption and desorption isotherms. Surface areas down to 0.01 $m^2 g^{-1}$ can be measured together with pore volume distributions. Any non-corrosive gas can be used.

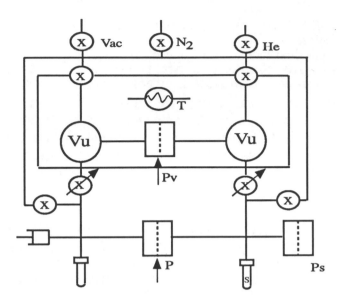

Fig. 2.10 Diagram of the Gemini Analyzer showing the connection of the two matched tubes by the servo valve mechanism.

(i) Micromeretics Digisorb

The Digisorb 2600 performs fully automated determinations of surface area, pore volume, pore size and pore area using a variety of gases with nitrogen and krypton as standard. The instrument can operate for four days completely unattended

(j) Micromeretics Gemini

The Micromeretics Gemini consists of two tubes of matched internal volume, one of which contains the sample while the other one is empty [193]. These two tubes are joined by servo valves as shown in Figure 2.10. With a previously degassed sample in position, a vacuum is pulled on the manifold system to expel residual gas before opening the tubes to the system. When a sufficient vacuum is obtained, the valves to the tubes are opened and the system brought into volumetric balance with an adjusting piston incorporated into the manifold. The sample and balance tubes are immersed in the coolant bath and, when thermal equilibrium has been obtained, the adsorbate gas is introduced into the manifold at the first desired relative pressure. This gas has equal access to both the sample and the balance tube. Because there is no sample in the balance tube, no adsorption of the gas takes place whereas, in the sample tube, gas is adsorbed on to the powder surface. As the sample adsorbs gas, the pressure in the sample tube falls and this unbalances the manifold and servo valve circuit. This causes more gas to be brought into the manifold to rebalance the pressures, the amount of gas being the same as that adsorbed on to the powder. This continues until the highest desired relative pressure is reached. Any deviation from ideal gas behavior is compensated for by this pressure balance system and the same applies to thermal gradients in the coolant bath. A typical five point analysis requires about 10 min, as opposed to an hour with conventional equipment, with no loss in accuracy.

The instrument is available as the Gemini 2360 and the Gemini 2370, the latter being the more sophisticatedd version with the capability of carrying out micropore analyses by *t*-plot.

(k) Quantachrome Nova 1000 and Nova 1200

The Nova 1000 operates without the need for dead space determination, thus obviating the need for helium. It can generate a single BET analysis, a multiple BET analysis, and a 25 point adsorption and desorption isotherm together with total pore volume and sample density. The user places the sample in a calibrated sample cell and after outgassing transfers it to one of the two measurement ports. An optional five port degassing station is also available. The nitrogen adsorbate may be taken from a gas cylinder or from a Dewar flask.

The Nova 1200 can use any non-condensing, non-corrosive gas for sorption analysis. This, coupled with a Microsoft Windows data reduction package significantly broadens the Nova's analytical range. A key application is in the characterization of microporous materials for which the BET equation fails. The most commonly used gas for this procedure is carbon dioxide at ice-water temperature. The Nova 1200 automatically measures detailed adsorption and desorption isotherms which are then analyzed using the Dubinin–Radushkevich theory to give micropore volume, surface area, average micropore width and adsorption energy.

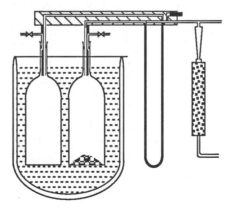

Fig. 2.11 The Ströhline Areameter.

(l) Ströhline Areameter

The Ströhline Areameter (Figure 2.11) is a simple, single point apparatus. Degassing is carried out in a heating thermostat that is capable of degassing eight samples simultaneously [194,195]. The adsorption vessel, containing the sample, together with a similar reference vessel is filled with nitrogen at atmospheric pressure. The two vessels are immersed in liquid nitrogen and the nitrogen adsorbed by the sample leads to a pressure difference between the two vessels which is measured on a differential oil manometer. The amount of nitrogen adsorbed by the sample is calculated from the pressure difference and the atmospheric pressure.

The conventional single point apparatus determines the amount adsorbed at a fixed relative pressure of either 0.2 or 0.3 according to the manufacturer. It is assumed that the BET line passes through the origin and this introduces an error, the magnitude of which depends on the BET c value (see section 2.4.1). The multipoint instruments can be manually driven or automatic and often operate with a wide range of adsorbates.

2.21.2 Continuous flow gas chromatographic methods

This method is a modification of gas adsorption chromatography in which the column packing is the sample itself and the mobile gas phase is a mixture of a suitable adsorbate and an inert gas.

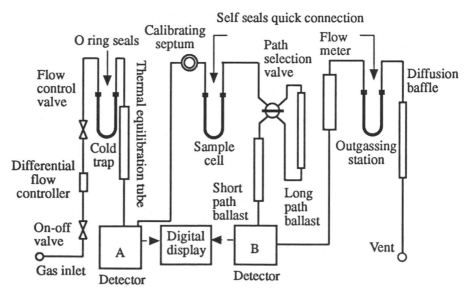

Fig. 2.12 Schematic diagram of the Quantachrome continuous flow gas chromatographic apparatus.

Table 2.5 Commercial gas adsorption surface area and pore size determination analyzers

Manufacturer and Model	Mode and Principle	Gas	Detection	Range $(m^2\,g^{-1})$	Ports	Data points	Sorption type
Bel, Japan Belsorp 28SA	Automatic static	Organic	Volume	0.5+	3	200	Physical
Bel, Japan Belsorp 18	Automatic static	Water+	Volumetric	1+	1	Multiple	Physical
Beta Scientific 4200	Automatic flow	N_2, He	Therm. cond.	0.1>2000	1	1 or 3	Physical
Beta Scientific 4201	Automatic flow	N_2, He	Therm. cond.	0.1>2000	1	1 or 3	Physical
Beta Scientific 4202	Automatic flow	$N_{2}2$, He	Therm. cond.	0.1>2000	2	1 or 3	Physical
Beta Scientific 4203	Automatic flow	N_2, He	Therm. cond.	0.1>2000	3	1 or 3	Physical
Carlo Erba Sorpty 1750	Manual static	N_2	Volume	1-1000	1	1	Phys./Chem.
Chandler Ni-Count-1	Manual static	N_2	Diff. press.	>0.1	1	1	Physical
Coulter Omnisorp 100	Automatic static	Non-	Diff. press.		1	<2000	Physical
Coulter Omnisorp 200	Automatic static	corros.	Diff. press.		3	<2000	Physical
Coulter Omnisorp 610	Automatic static		Diff. press.		10	<2000	Physical
Coulter SA 3100	Automatic flow		Therm. cond.		3	Multiple	Physical
Fisons Sorptomatic 1800	Automatic static	N_2, Ar	Volume	>0.2	1	Multiple	Physical
Fisons Sorptomatic 1900	Automatic static	Multi	Diff. press.	>0.005	2	Multiple	Phys./Chem.
Horiba SA-6200 Series	Automatic flow	N_2	Therm. cond.	>0.1	1–3	Multiple	Physical
Leeds & Northrop 4200	Manual flow	N_2, He	Therm. cond.	>0.5	1	1	Physical
Leeds & Northrop 4201	Automatic flow	N_2, He	Therm. cond.	0.15-4000	1	1	Physical
Micromeretics 2200	Manual static	N_2	Volume	0.5-1000	1	1	Physical
Micromeretics 2300	Manual flow	N_2, He	Therm. cond.	0.01-1000	1	Multiple	Physical
Micromeretics ASAP 2000	Automated flow	Multi	Volume		1	Multiple	Physical
Micromeretics ASAP 2400	Automatic static	N_2	Volume	0.01-1000	6	Multiple	Physical
Micromeretics Gemini 2360	Manual static	N_2	Volume	0.1-100	1	Multiple	Physical
Micromeretics Gemini 2370	Manual static	N_2	Volume	0.1-100	1	Multiple	Physical

Table 2.5 (Cont.) Commercial gas adsorption surface area and pore size determination analyzers

Manufacturer and Model	Mode and Principle	Gas	Detection	Range ($m^2\,g^{-1}$)	Ports	Data points	Sorption type
Micromeretics Digisorb 2600	Automatic static	N_2	Volume	0.01-10^3	5	Multiple	Physical
Micromeretics Accusorb 2100E	Manual static	N_2	Volume	0.01-10^3	4	Multiple	Phys./Chem.
Micromeretics Chemisorb 2700	Manual flow	H_2,O_2	Therm. cond.	0.02-280	2	Multiple	Chemical
Micromeretics Chemisorb 2800	Automatic pulsed	CO, SO_2 varied					Phys./Chem.
Micromeretics TPD/TPR 2900	Automatic flow	N_2	Therm. cond.		1	Multiple	Physical
Micromeretics Flowsorb II 2300	Automatic flow	N_2	Therm. cond.		1	Multiple	Physical
Netsch Gravimat	Automatic static	N_2, Kr	Gravimetric	>0.1	1	Multiple	Physical
Thermo-Gravimat	Automatic static	N_2, Kr	Gravimetric	>0.1	1	Multiple	Physical
Omicron Omnisorb 100C	Automatic either	H_2,CO	Gravimetric	>0.01	1	Multiple	Chemical
Omicron Omnisorb 360	Automatic flow	N_2	Volume	>0.01	4	Multiple	Phys./Chem.
Quantachrome Monosorb	Manual flow	Any	Therm. cond.	0.1-10^{3+}	1	1	Phys./Chem.
Quantachrome Quantasorb	Manual flow	Any	Therm. cond.	>0.1	1	Multiple	Phys./Chem.
Quantachrome Quantasorb Jr	Manual flow	Any	Therm. cond.	>0.1	1	Multiple	Phys./Chem.
Quantachrome Autosorb I	Automatic static	Any	Volume	>0.1	1	Multiple	Phys./Chem.
Quantachrome Autosorb 6B	Automatic static	N_2	Volume	>0.1	6	Multiple	Phys./Chem.
Quantachrome Nova 1000	Automatic static	Any	Volume	0.01-10^3	2	Multiple	Physical
Quantachrome Nova 1200	Automatic static	Any	Volume	0.01-10^3	2	Multiple	Physical
Strohline Areameter	Manual static	N_2	Diff. press.	>0.1	1	1	Physical
Strohline Area Mat	Automatic static	N_2	Diff. press.	>0.1	1	1	Physical
Den-Ar-Mat	Automatic static	N_2	Volume	>0.1	1	Multiple	Physical
Surface Measurement Systems (DVS)	Automatic static	HN_2O	Mass		1	Multiple	Physical

A known mixture of the gases, often nitrogen and helium, is passed through a thermal conductivity cell, through the sample and then to a recording potentiometer via the thermal conductivity cell. When the sample is immersed in liquid nitrogen it absorbs nitrogen from the mobile phase. This unbalances the thermal conductivity cell, and a pulse is generated on a recorder chart. Removing the coolant gives a desorption peak equal in area and in the opposite direction to the adsorption peak. Since this is better defined than the adsorption peak it is the one used for surface area determination.

Calibration is effected by injecting sufficient air into the system to give a peak of similar magnitude to the desorption peak and obtaining the ratio of gas adsorbed per unit peak area. (Air can be used instead of nitrogen since it has the same thermal conductivity.)

A schematic of the apparatus is shown in Figure 2.12. A nitrogen/helium mixture is used for a single point determination and multiple points can be obtained using several such mixtures or premixing two streams of gas. Calculation is essentially the same as for the static method but no dead space correction is required.

(a) Beta Scientific

Beta Scientific make three analyzers, the 4200, the 4201 and the 4203. These instruments were designed for unattended and fully automatic operation, complete analyses in less than 10 minutes, cost effectiveness and an ASCII, R232 interface for computer or LIMS connection [196]. The model 4200 is a single channel instrument with no R232 interface and manual calibration; the model 4201 also includes an R232 interface, and the moel 4203 is a three channel version without sample preparation but with an R232 interface. The model 4210 is a three station sample preparation unit, a required accessory for model 4203.

(b) Leeds & Northrup Model 4200

This is a single port, single point surface area analyzer designed to generate a surface area in less than 5 minutes. Samples are degassed in a separate four sample preparation facility. Once it has been prepared it is transferred to the test station and the operation is completed automatically. In addition the analyzer is capable of multipoint analyses with manual selection of gas mixtures.

(c) Micromeretics Rapid surface area analyzer 2300

This instrument measures BET surface area in minutes using a preset single point method. Three sample ports are provided.

(d) Micromeretics Flowsorb II 2300

This is a low price instrument capable of single-point, multipoint and total pore volume analysis using pre-mixed gases or a gas mixer. Automatic operation is available with optional Automate 23.

(e) Quantachrome Monosorb

This is a fully automated, single point, instrument operating in the surface area range 0.1 to 250 m^2 with a measurement time of around 6 min and a reproducibility better than 0.5%. Its autocalibrate feature eliminates the need for calibrating each sample. Degassing is carried out in an in-built degassing station.

(f) Quantachrome Quantasorb

This is a manual multipoint instrument in which a variety of gases can be used as adsorbates. For rapid operation pre-mixed gases can be used and if many data points are required a linear mass flowmeter is available to dial in the flow rates of the adsorbate and the carrier gas. The area under the desorption peak is determined automatically using a built-in digital integrator.

(g) Quantachrome Quantasorb Jr

The Quantachrom Jr is a low cost manual instrument designed for surface area, pore size distribution and chemisorption studies.

2.21.3 Gravimetric methods

(a) Surface Measurement Systems Dynamic Vapor Sorption (DVS) Analyzer

In the DVS the sample is placed on a microbalance which is exposed to a continuous flow of air of known humidity. An ultrasensitive Cahn microbalance allows vapor sorption measurements on sample sizes as small as 1 mg to be analyzed with a resolution of 0.1 µg or as large as 100g with a resolution of 10 µg.

(b) Netzch Gravimat

The Netzsch Gravimat, based on a design by Robens and Sandstede [197], is equipped with one or more electromagnetic micro-balances. An additional micro-balance is used to measure the pressure and counterbalance the buoyancy effect. A turbomolecular pump permits evacuation down to a pressure of 10^{-5} Pa. The gas pressure can be varied in 100 steps from 1 to 2 x 10^5 Pa. The degassing temperature can be varied up to 2000 K and the measurement temperature down to 77 K.

References

1 Brunauer, S., Deming, L.S., Deming, W.S. and Teller, E. (1940), *J. Am. Chem. Soc.*, **62**, 1723, *40*
2 Joyner, L.G., Weinberger, E.B. and Montgomery, C.W. (1945), *J. Am. Chem. Soc.*, **67**, 2182–2188, *40*
3 Brunauer, S. and Emmett, P.H. (1937), *J. Am. Chem. Soc.*, **59**, 2682, *41*
4 Emmett, P.H. and Dewitt, T.W. (1937), *J. Am. Chem. Soc.*, **59**, 2682, *41*
5 Wade, W.H. and Blake, T.D. (1971), *J. Phys. Chem.*, **75**, 1887, *42*
6 Emmett, P.H. and Brunauer, S. (1937), *J. Am. Chem. Soc.*, **59**, 1553, *42*
7 Orr, C. and Dallavalle, J.M. (1959), *Fine Particle Measurement*, Macmillan, N.Y., *42, 51, 64, 79*
8 Adamson, A.W. (1960), *Physical Chemistry of Surfaces*, Interscience, N.Y., *42, 43, 64*
9 Parfitt, G.D. and Sing, K.S.W. eds, (1971), *Characterisation of Powder Surfaces*, Academic Press, *42*
10 Boer, J.H. de, Kaspersma, J.H. and Dongen, R.H. van (1972), *J. Colloid Interf. Sci.*, **38**(1), 97–100, *42*
11 Singleton, J.H. and Halsey, G.D. (1954), *J. Phys. Chem.*, **58**, 1011, *42*
12 Boer, J.H. de *et al.* (1967), *J. Catalysis*, **7**, 135–139, *42*
13 Langmuir, I. (1918), *J. Am. Chem. Soc.*, **40**, 1368, *43*
14 Volmer, M. (1925), *Z. Phys. Chem.*, **115**, 253, *45*
15 Fowler R.H. (1935), *Proc. Camb. Phil. Soc.*, **31**, 260, *45*
16 Young, D.M. and Crowell, A.D, (1962), *Physical Adsorption of Gases*, p. 104, Butterworths, *45, 51, 52, 62, 64, 81*
17 Halsey, G.D. and Taylor, H.S. (1947), *J. Chem. Phys.*, **15**(9), 624–630, *46*
18 Burevski, D. (1975), PhD thesis, University of Bradford, U.K., *46*
19 Halsey, G. and Taylor, H.S. (1947), *J. Chem. Phys.*, **15**(9), 624–630, *46*
20 Sips, R. (1948), *J. Chem Phys.*, **16**, 490, *46*
21 Sips, R. (1950), *J. Chem Phys.*, **18**, 1024, *46*
22 Brunauer, S., Emmett, P.H. and Teller, E. (1938), *J. Am. Chem. Soc.*, **60**, 309, *47*
23 Baley, E.G.G. (1937), *Proc. Royal Soc.*, A160, 465, *48*
24 Brunauer, S., Emmett, P.H. and Teller, E. (1938), *J. Am. Chem. Soc.*, **60**, 309, *48*
25 Corrin, M.L. (1953), *J. Am. Chem. Soc.*, **75**, 4623, *51*
26 Loeser, E.H., Harkins, W.D. and Twiss, S.B. (1953), *J. Phys. Chem.*, **57**, 591, *51*

27   MacIver, D.S. and Emmett, P.H. (1953), *J. Phys. Chem.*, **60**, 824, *51*
28   Brunauer, S. *et al.* (1940), *J. Am. Chem.* Soc., **62**, 1723, *52*
29   Cassel, H.M. (1944), *J. Chem. Phys.*, **12**, 115, *52*
30   Cassel, H.M. (1944), *J. Phys. Chem.*, **48**, 195, *52*
31   Halsey, G.D. (1948), *J. Chem. Phys.*, **16**, 931, *52*
32   Gregg, S.J. and Jacobs, J. (1948), *Trans. Faraday Soc.*, **44**, 574, *53*
33   Halsey, G.D. (1948), *J. Chem. Phys.*, **16**, 931, *53*
34   Dollimore, D., Spooner, P. and Turner, A. (1976), *Surface Technol.*, **4**(2), 121–160, *53*
35   Gregg, S.J. and Jacobs, J. (1948), *Trans. Faraday Soc.*, **44**, 574, *53*
36   Genot, B. (1975), *J. Colloid Interf. Sci.*, **50**(3), 413–418, *54*
37   Hill, T.L. (1946), *J. Chem. Phys.*, **14**, 268, *55*
38   Lowell, S. (1975), *Powder Technol.*, **2**, 291–293, *55*
39   Lowell, S. and Shields, J.E. (1991), *Powder Surface Area and Porosity*, p. 25, 2nd ed, Chapman & Hall, *56*
40   Jones, D.C. and Birks, E.W. (1950), *J. Chem. Soc.*, 1127, *56*
41   Jones, D.C. (1951), *J. Chem. Soc.*, 1461, *56*
42   Brunauer, S. (1970), *Surface Area Determination*, Proc. Conf. Soc. Chem. Ind., Bristol, 1969, Butterworths, *56*
43   Joyner, L.G., Weinberger, E.B. and Montgomery, C.W. (1945), *J. Am. Chem. Soc.*, **67**, 2182–2188, *57*
44   Gregg, S.J. and Sing, K.S.W. (1967), *Adsorption, Surface Area and Porosity*, Academic Press, N.Y., *57*
45   Pickett, G. (1945), *J. Am. Chem. Soc.*, **67**, 1958, *57*
46   Anderson, R.B. (1946), *J. Am. Chem. Soc.*, **68**, 686, *57*
47   Anderson, R.B. and Hall, W.K. (1948), *J. Am. Chem. Soc.*, **70**, 1727, *57*
48   Keenan, A.G. (1948), *J. Am. Chem. Soc.*, **70**, 3947, *57*
49   Brunauer, S. Skalny, J. and Bodor, E.E. (1969), *J. Colloid Interf. Sci.*, **30**, 546, *58*
50   Barrer, R.M., MacKenzie, N. and McLeod, D. (1952), *J. Chem. Soc.*, 1736, *58*
51   Polanyi, M. (1914), *Verh. Deutsch Phys. Ges.*, **16**, 1012, *58*
52   Niemark, A.V. (1994), *J. Colloid Interf. Sci.*, **165**, 91–96, *58*
53   Dubinen, M.M. and Radushkevich, L.V. (1947), *Dokl. Akad. Nauk.*, *SSSR*, **55**, 331, *58*
54   Radushkevich, L.V. (1949), *Zh. Fiz. Khim.*, **23**, 1410. *58*
55   Dubinen, M.M. (1960), *Chem. Rev.*, **60**, 235, *59*
56   Dubinen, M.M. and Timofeev, D.P. (1948), *Zh. Fiz. Khim.*, **22**, 133, *59*
57   Nikolayev, K.M. and Dubinen, M.M. (1958), *Izv. Akad .Nauk.*, *SSSR, Otd. Tekn. Nauk.*, 1165, *59*
58   Dubinen, M.M., Niemark, A.V. and Serpinsky, V.V. (1992), *Commun. Akad. Sci.*, USSR, Chem., **1**(13), *59*

59   Dubinen, M.M., Niemark, A.V. and Serpinsky, V.V. (1993), *Carbon*, **31**, 1015, *59*
60   Lamond, T.G. and Marsh, H. (1963), *Carbon*, **1**, 293, *59*
61   Walker, P.L. and Shelef, M. (1967), *Carbon*, **5**, 7, *59*
62   Freeman, E.M. *et al.*, (1970), *Carbon*, **8**, 7, *59*
63   Toda, Y. *et al*, (1970), *Carbon*, **8**, 565, *59*
64   Freeman, E.M. *et al.*, (1970), *Carbon*, **8**,7, *59*
65   Dubinen, M.M. (1972), *Adsorption–Desorption Phenomena*, Academic Press, p. 3, *59*
66   Dovaston, N.G., McEnany, B. and Weeden, C.J. (1972), *Carbon*, **10**, 277, *59*
67   Bering, B.P., Dubinen, M.M. and Serpinsky, V.V. (1966), *J. Colloid Interf. Sci.*, **21**, 378, *59*
68   Bering, B.P., Dubinen, M.M. and Serpinsky, V.V. (1971), *Izv. Akad. Nauk.*, *SSSR, Ser Khim.* (English), **17**, *59*
69   Schram, A. (1965), *Nuava Cimento*, Suppl. 5, 309, *59*
70   Ricca, F., Medana, R. and Bellarda, A. (1967), *Z. Phys. Chem.*, **52**, 276, *59*
71   Hobson, J.P. and Armstrong, P.A. (1959), *J. Phys. Chem.*, **67**, 2000, *60*
72   Hobson, J.P. (1967), *The Solid–Gas Interface*, p. 14, ed. E.A. Flood,, Arnold, London, *60*
73   Kadlec, O. (1976), *Dechema Monogr.*, **79**, 1589–1615, 181–190, *60*
74   Rosai, L. and Giorgi, T.A. (1975), *J. Colloid Interf. Sci.*, **51**(2), 217–224, *60*
75   Stoekli, H.F. (1977), *J. Colloid Interf. Sci.*, **59**(1), 184–185, *60*
76   Marsh, H. and Rand, B. (1970), *J. Colloid Interf. Sci.*, **33**, 101–116, *60*
77   Marsh H. and Rand, B. (1971), *3rd Conf. Industrial Carbon and Graphite*, *Soc. Chem. Ind.*, p. 212, *60*
78   Rand, B. (1976), *J. Colloid Interf. Sci.*, **56**(2), 337–346, *60*
79   Marsh, H. and Siemiensiewska, T. (1965), *Fuel*, **44**, 335, *60*
80   Kaganer, M.G. (1959), *Zh. Fiz. Khim.*, **32**, 2209, *60*
81   Kaganer, M.G. (1959), *Russ. J. Phys. Chem.*, **33**, 352, *60*
82   Ramsey, J.D.F. and Avery, R.C. (1975), *J. Colloid Interf. Sci.*, **51**(1), 205–208, *61*
83   Jovanovic, D.S. (1969), *Kolloid, Z.Z. Poly.*, **235**, 1203, *61*
84   Jovanovic, D.S. (1969), *Kolloid, Z.Z. Poly.*, **235**, 1214, *62*
85   Hüttig G.F. (1948), *Monatsh. Chem.*, **78**, 177, *62*
86   Ross, S. (1953), *J. Phys. Chem.*, **53**, 383, *62*
87   Harkins, W.D. and Jura, G. (1943), *J. Chem. Phys.*, **11**, 430–431, *62*
88   Livingstone, H.K. (1944), *J. Chem. Phys.*, **12**, 466, *64*
89   Livingstone, H.K. (1947), *J. Chem. Phys.*, **15**, 617, *64*
90   Emmett, P.H. (1946), *J. Am. Chem. Soc.*, **68**, 1784, *64*
91   Smith, T.D. and Bell, R. (1948), *Nature*, **162**, 109, *64*

92   Hill, T.L. (1952), *Adv. Catal.*, **4**, 211, *64*
93   Parfitt, G.D. and Sing, K.S.W. (1971), eds, *Characterization of Powder Surfaces*, Academic Press, *65*
94   Alzamora, L. and Cortes, J. (1976), *J. Colloid Interf. Sci.*, **56**(2), 347–349, *65*
95   Oulton, T.D. (1948), *J. Phys. Colloid Chem.*, **52**, 1296, *65*
96   Schüll, C.G. (1948), *J. Am. Chem. Soc.*, **70**, 1405, *65*
97   Wheeler, A. (1955), *Catalysis*, Reinhold, N.Y., *65*
98   Lippens, B.C., Linsen., B.G. and de Boer, J.H. (1964), *J. Catalysis*, **3**, 32–37, *65*
99   Parfitt, G.D., Sing, K.S.W. and Unwin, D. (1975), *J. Colloid Interf. Sci.*, **53**(2), 187–193, *66*
100   Lecloux, A. (1970), *J. Catalysis*, **81**, 22, *66*
101   Mikhail, R. Sh., Guindy, N.M. and Ali, I.T. (1976), *J. Colloid Interf. Sci.*, **55**(2), 402–408, *66*
102   Pierce, C. (1968), *J. Phys. Chem.*, **72**, 3673, *66*
103   Lamond, T.G. and Marsh, H. (1964), *Carbon*, **1**, 281, 293, *66*
104   Marsh, H. and Rand, B. (1970), *J. Colloid Interf. Sci.*, **33**(3), 478–479, *66*
105   Kadlec, O. (1976), *Dechema Monogr.*, **79**, 1589–1615, 181–190, *66*
106   Dubinin, M.M. *et al* (1975), *Izv. Akad. Nauk. SSSR, Ser. Khim.*, **6**, 1232–1239, *66*
107   Kiselev, A.V. (1945), *USP Khim.*, **14**, 367, *67*
108   Brunauer, S. (1970), *Surface Area Determination*, Proc. Conf. Soc. Chem. Ind., Bristol, 1969, Butterworths, *67*
109   Brunauer, S. and Mikhail, R. Sh. (1975), *J. Colloid Interf. Sci.*, **523**(3), 187–193, *67*
110   Kadlec, O. (1976), *Dechema Monogr.*, **79**, 1589–1615, 181–190, *67*
111   Kadlec, O. (1969), *J. Colloid Interf. Sci.*, **31**, 479, *67*
112   Joy, A.S. (1953), *Vacuum*, 3, 254, *69, 80*
113   Emmett, P.H. (1940), 12th report of the committee on catalysis, *Physical Adsorption in the Study of the Catalysis Surfac*e, Ch. 4,Wiley, N.Y., *69*
114   Emmett, P.H. (1944), *Colloid Chemistry*, V. Reinhold, N.Y., *69*
115   Harvey, E.N. (1947), *ASTM Symp., Paint and Paint Material*, *69*
116   Joyner, L.G. (1949), *Scientific and Industrial Glass Blowing and Laboratory Techniques*, Instruments Publ. Co., Pittsburgh, PA, U.S.A., *69*
117   Schubert, Y. and Koppelman, B. (1952), *Powder Metal. Bull.*, **6**, 105, *69*
118   Vance, R.F. and Pattison, J.N. (1954), *Special Report on Apparatus for Surface Area Determination and other Adsorption Studies on Solids*, Battelle Memorial Inst., Ohio, U.S.A., *69*
119   Harkins, W.D. and Jura, G. (1944), *J. Am. Chem. Soc.*, **66**, 1366, *69*

120 Thompson, J.B., Washburn, E.R. and Guildner, L.A. (1952), *J. Phys. Chem.*, **56**, 979, *69*

121 Bensen, S.W. and Ellis, D.A. (1948), *J. Am. Chem. Soc.*, **70**, 3563, *69*

122 Dollimore, D., Rickett, G. and Robinson, R. (1973), *J. Phys. E.*, **6**, 94, *69*

123 Bugge, P.E. and Kerlogue, R.H. (1947), *J. Soc. Chem. Ind.*, London, **66**, 377, *69*

124 Lippens, B.C., Linsen, B.G. and de Boer, J.H. (1964), *J. Catalysis*, **3**, 32–57, *69*

125 Rasneur, B. and Charpin, J. (1975), *Fine Particle Int. Conf.*, ed. W.H. Kuhn, Electrochem. Soc. Inc., Princeton, N.J., *69*

126 Jaycock, M.J. (1981), *Particle Size Analysis*, 87–101, edS. N.G. Stanley–Wood and T. Allen, Proc. Particle Size Analysis Conf., Royal Soc. Chem., Analyt. Div., publ. Wiley–Heyden, *70*

127 BS 4359 (1984), *Determination of the Specific Surface Area of Powders*, Part 1. Recommendations for gas adsorption techniques, *70*

128 Harris, M.R. and Sing, K.S.W. (1955), *J. Appl. Chem.*, **5**, 223, *70*

129 Lippens, B.C., Linsen, B.G. and de Boer, J.H. (1964), *J. Catalysis*, **3**, 32, *70*

130 Pickering, H.L. and Eckstrom, H.C. (1952), *J. Am. Chem. Soc.*, **70**, 3563, *70*

131 Tomlinson, L. (1954), *UKAEA Report* 1, G.R.–TN/S–1032, *70*

132 Haul, R.A.W. (1956), *Angew Chem.*, **68**, 238, *70*

133 Carden, J.L. and Pierotti, R.A. (1974), *J. Colloid Interf. Sci.*, **47**(2), 379–394, *70*

134 Wooten, L.A. and Brown, C. (1943), *J. Am. Chem. Soc.*, **65**, 113, *70*

135 Lister. B.A.J. and McDonald, L.A. (1952), *UK AERE Report*, C/R 915, *70*

136 Aylmore, D.W. and Jepson, W.B. (1961), *J. Scient. Instrum.*, **38**(4), 156, *70*

137 Choudhary, V.R. (1974), *J. Sci. Ind. Res.*, **33**(12), 634–641, *71*

138 Cahn, L. and Schultz, H.R. (1962), *Vac. Microbalance Technol.*, **3**, 29, *72*

139 McBain, J.W. and Bakr, A.M. (1926), *J. Am. Chem. Soc.*, **48**, 690, *69, 72, 79*

140 Morris, H.E. and Maass, P. (1933), *Can. J. Res.*, **9**, 240, *72*

141 Dunn, R.C. and Pomeroy, H.H. (1947), *J. Phys. Colloid Chem.*, **51**, 981, *72*

142 McBain, J.W. and Britton, H.T.S. (1930), *J. Am. Chem. Soc.*, **52**, 2198, *72*

143 McBain, J.W. and Sessions, R.F. (1948), *J. Colloid Sci.*, **3**, 213, *72*

144 Boyd, G.E. and Livingstone, H.K. (1942), *J. Am. Chem. Soc.*, **64**, 2838, *72*

145   Seborg, C.O., Simmons, F.A. and Baird, P.K. (1936), *Ind. Eng. Chem., Ind. Edn.*, **28**, 1245, 72

146   Dubinin, M.M. and Timofeev, D.P. (1947), *Zh. Fiz. Khim.*, **21**, 1213, 72

147   Lemke, W. and Hofmann, U. (1934), *Angew Chem.*, **47**, 37, 72

148   Seborg, C.O. and Stamm, A.J. (1931), *Ind. Eng. Chem., Ind. Edn.*, **23**, 1271, 73

149   Pidgeon, L.M. and Maass, O. (1950), *J. Am. Chem. Soc.*, **52**, 1053, 73

150   Mulligan, W.O. *et al.*, (1951), *Anal. Chem.*, **23**, 739, 73

151   Stamm, A.J. and Woodruff, S.A. (1941), *Ind. Eng. Chem., Analyt. Edn.*, **13**, 386, 73

152   Barrett, H.M., Birnie, A.W. and Cohen, M. (1940), *J. Am. Chem. Soc.*, **62**, 2839, 73

153   Gulbransen, E.A. (1944), *Rev. Sci. Instrum.*, **15**, 201, 73

154   Rhodin, T.N. (1950), *J. Am. Chem. Soc.*, **72**, 4343, 73

155   Rhodin, T.N. (1950), *J. Am. Chem. Soc.*, 5691, 73

156   Rhodin, T.N. (1951), *Adv. Catalysis*, **5**, 39, 73

157   Bowers, R. and Long, E.A. (1955), *Rev. Sci. Instrum.*, **23**, 259, 73

158   Gregg, S.J. (1946), *J. Chem. Soc.*, 561, 564, 74

159   Gregg, S.J. and Wintle, M.F. (1946), *J. Sci. Instrum.*, **23**, 259, 74

160   Cahn, L. and Schultz, H.R. (1962), *Vac. Microbalance Technol.*, **3**, 29, 74

161   Lobenstein, W.V. and Dietz, V.R. (1951), *J. Res. Natl. Bur. Stand.*, **46**, 51, 74

162   Nelson, F.M. and Eggertsen, F.T. (1958), *Anal. Chem.*, **30**, 1387, 74

163   Ellis, J.F., Forrest, C. W. and Howe, D.D. (1960), *UK AERE Report D–E, G.R. 229* (CA), 76

164   Atkins, J.H. (1964), *Anal. Chem.*, **36**, 579, 76, 81

165   Haley, A.J. (1963) *J. Appl. Chem.*, **13**, 392, 76

166   Lapointe, C.M. (1970), *Can. Mines Br. Tech. Bull.*, TB119, 27, 77

167   Lowell, G.H.B. (1970), *Surface Area Determination*, Proc. Conf. Soc. Chem. Ind., Bristol, Butterworths, 77

168   Tucker, B.G. (1947), *Anal. Chem.*, **47**(4), 78–79, 78

169   Bhat, R.K. and Krishnamoorthy, T.S. (1976), *Ind. J. Technol.*, **14**(4), 170–171, 78

170   Yen, C–M., and Chang, C–Y. (1977), *Hua Hsueh Pao*, **35**(3–4), 131–140, 78

17`   Emmett, P.H. (1941), *ASTM Symp. New Methods of Particle Size Determination in the Sub–Sieve Range*, pp. 95–105, 76

172   Bugge, P.E. and Kerlogue, R.H. (1947), *J. Soc. Chem. Ind.*, London, 66, 377, 76

173   Holmes, H.F., Fuller, E.L. Jr and Beh, R.A. (1974), *J. Colloid Interf. Sci.*, **47**(2), 365–371, 79

174   Lopez–Gonzales, J. de D., Carpenter, F.G. and Dietz, V.R. (1955), *J. Res. Nat. Bur. Stand.*, **55**,11, 79

175 Beebe, R.A., Beckwith, J.B. and Honig, J.M. (1945), *J. Am. Chem. Soc.*, **67**, 1554, *80*
176 Singleton, J.H. and Halsey, G.D. (1954), *J. Phys. Chem.*, **58**, 1011, *80*
177 Boer, J.H. de., *et al.* (1967), *J. Catalysis*, 7, 135–139, *80*
178 Sing, K.S.W. and Swallow, D. (1960), *J. Appl. Chem.*, 10, 171, *80*
179 Zettlemoyer, A.C., Chand, A. and Gamble, E. (1950), *J. Am. Chem. Soc.*, **72**, 2752 243, *80*
180 Jaycock, M.J. (1977), *The Krypton BET Method,* Chemistry Dept., Univ. Technol., Loughborough, Leics., LE11 3TU, U.K., *80*
181 Litvan, G.G. (1972), *J. Phys. Chem.*, 76, 2584, *80*
182 Knowles, A.J. and Moffatt, J.B. (1972), *J. Colloid Interf. Sci.*, **41**(1), 116–123, *80*
183 Dzisko, V.A. and Krasnopolskaya, V.N. (1943), *Zh. Fiz. Khim.*, **26**, 1841, *81*
184 Pierce, C. (1968), *J. Phys. Chem.*, **72**, 3673, *81*
185 Baker, F.S and Sing, K.S.W. (1976), *J. Colloid Interf. Sci.*, **55**(3), 605–613, *81*
186 Avnir, D. (1989), *The Fractal Approach to Heterogeneous Chemistry*, John Wiley & Sons , *81*
187 Wooten, L.A. and Brown, C. (1943), *J. Am. Chem. Soc.*, **65**, 113, *81*
188 Lister, B.A.J. and MacDonald, L.A. (1952), *U.K. AERE Report* C/R 915, *81*
189 Desbiens, G. and Zwicker, J.D. (1976), *Powder Technol.*, **13**, 15–21, *81*
190 Anon. (1976), *Measure of Powder Fineness Resulting from Inter–laboratory Studies,* Assoc. France Normalisation, Tour Europe, Cedex–7–92080, Paris, La Defense, *81*
191 Haynes, J.M. (1959), *Clay Minerals Bull.*, **4**(22), 69–74, *86*
192 Kouichi, M., Hayashi, J. and Hashimoto, K. (1991), *Carbon* **29**(4/5), 653–660, *88*
193 Camp, R.W. and Stanley, H.D. (1991), *American Laboratory*, September, 34–35, *90*
194 Haul, R.A.W. and Dümbgen, G. (1960), *Chem. Ingr. Tech.*, Part 1, **32**, 349, *92*
195 Gaul, L. (1964), *Angew Mess., Regeltech.*, 4(12), 107–111, *92*
196 Hershenhart, E.J., Zadnik, V.C., Zemia, W. and Jennings, H. (1990), *Am. Laboratory,* 84–90, Jan., *95*
197 Robens, E. and Sandstede, G. (1969), *J. Sci. Instrum.*, **2**, 365, *96*

3

Determination of pore size distribution by gas adsorption

3.1 Introduction

Pore size and size distribution have significant effects over a wide range of phenomena from the absorbency of fine powders in chemical catalysis to the frost resistance of bricks. To investigate these effects, pore size measurements have been described using a wide range of techniques and apparatus. Pore surface area is generally accepted as the difference between the area of the surface envelope of the particle (i.e. superficial area) and its total surface area. The pores may be made up of fissures and cavities in the particle; they may have narrow entrances and wide bodies (ink bottle pores) or they may be V-shaped. In order that their magnitude and distribution may be determined, it is necessary that they be accessible to the measuring fluid, i.e. they must be open pores.

The presence and extent of open pores may be found by immersing the powder in mercury and measuring the liquid displacement, then repeating the exercise using a helium pyknometer. Since mercury does not wet most solids it leaves the pores unfilled and the difference between the two volumes is the pore volume [1]. Closed pores may be evaluated by grinding the powder, which opens up some of these pores, thus decreasing the apparent solid volume [2]. Total pore volume may be determined by boiling the powder in a liquid, decanting the liquid and determining the volume of liquid taken up by the solid after it has been superficially dried [2]. Pore size distribution may be found by using a range of liquids of different molecular sizes [3–6]. Direct visual examination under optical and electron microscopes has also been used. Porosity has also been measured using the absorption of gamma radiation [7].

Gas adsorption is widely used for pore size distribution determination. This is accomplished by measuring the volume of gas adsorbed or desorbed, at constant temperature and over a wide range of pressures, to generate adsorption and desorption isotherms. Manual determination of adsorption and desorption isotherms is tedious and time consuming and, for routine use, automatic analyzers are preferred.

These are capable of generating single or multiple analyses overnight and can be programmed to give BET surface area, together with graphical and tabular data on pore size distribution. Although distributions measured by gas adsorption cover a narrower pore diameter range than mercury porosimetry (2.0 nm to 400 nm cf. 3.0 nm to 1.6 mm) it is more widely used.

In order to calculate a pore size distribution, a model must be selected, e.g. cylindrical pores, wedge shaped pores, ink bottle pores and so on. It is also necessary to decide which branch of the isotherm to employ, adsorption or desorption. During desorption the cores of the pores are emptied leaving a residual layer, the thickness of which has to be known, in order to calculate a pore size distribution. All of these variables affect the derived distribution.

The theory for adsorption of vapor on to a porous solid is derived, from thermodynamic considerations, and leads to the Kelvin equation which is exact in the limit for large pores. However it becomes progressively less accurate as the pore size decreases and breaks down when the pore size is so small that the molecular texture of the fluid becomes important.

Although nitrogen adsorption isotherms are readily determined with high precision, the extraction of pore size distributions from the experimental data is problematical with small pores, due to the restricted range of validity of the Kelvin equation and the difficulty of assigning a correct value to the residual thickness, t, when the 'core' of a pore empties.

Below a critical size, pores do not undergo capillary condensation, but fill continuously as the pressure is increased without a discontinuity in the single pore adsorption isotherm [8].

For nitrogen adsorbing on porous carbon this critical pore size corresponds roughly to the conventional boundary between micropores and mesopores at 2nm [9]. The pore filling mechanism is not accounted for in the thermodynamic methods, which are therefore incapable of determining pore size distributions in the micropore range.

3.2 The Kelvin equation

The Kelvin equation may be derived as follows. Consider a liquid within a pore in equilibrium with its vapor. Let a small quantity, ∂a moles, be distilled from the bulk of liquid outside the pore, where its equilibrium pressure is P_0, into the pore where its equilibrium pressure is P. The total increase in free energy δG is the sum of three parts: evaporation of δa moles of liquid at pressure P_0 (δG_1); expansion of ∂a moles of vapor from pressure P_0 to pressure P (δG_2); condensation of δa moles of vapor to liquid at pressure P (δG_3).

Since condensation and evaporation are equilibrium processes $\delta G_1 = \delta G_3 = 0$ whilst the change in free energy during expansion is given by:

$$\delta G_2 = RT \ln\left(\frac{P}{P_0}\right)\delta a \tag{3.1}$$

the vapor being assumed to behave like a perfect gas.

The condensation of the vapor in the pores results in a decrease in the area of solid–liquid interface and an increase in solid–vapor interface (δS). The change in free energy during this process is:

$$\delta G' = -\delta S(\gamma_{SL} - \gamma_{SV}) \tag{3.2}$$

where

$$\gamma_{SL} - \gamma_{SV} = \gamma_{LV}\cos(\theta) \tag{3.3}$$

γ is the interfacial surface tension, suffixes referring to solid–liquid (SL), solid–vapor (SV) and liquid–vapor (LV); θ is the wetting angle which is taken as zero. Since

$$\delta G' = \delta G_2 \tag{3.4}$$

$$RT \ln(\frac{P}{P_0})\delta a = -\gamma_{LV}\cos(\theta)\delta S \tag{3.5}$$

The volume condensed in the pores is: $\delta V_c = V_L \delta a$ where V_L is the molar volume. Therefore:

$$\frac{\delta V_c}{V_L} RT \ln(\frac{P}{P_0}) = -\gamma_{LV}\cos(\theta)\delta S \tag{3.6}$$

The limiting case being:

$$\frac{dV_c}{dS} = \frac{-V_L \gamma_{LV}\cos\theta}{RT \ln\dfrac{P}{P_0}} \tag{3.7}$$

For cylindrical pores of radius r and length L:

$$V_c = \pi r^2 L \tag{3.8}$$

$$S = 2\pi r L \tag{3.9}$$

Hence

$$\frac{V_c}{S} = \frac{r}{2} \tag{3.10}$$

Equation (3.7) may therefore be written:

$$RT \ln x = \frac{-2\gamma_{LV}V_L \cos\theta}{r} \tag{3.11}$$

where $x = P/P_0$

For non–cylindrical pores, having mutually perpendicular radii r_1 and r_2, equation (3.11) becomes:

$$RT \ln x = -\gamma_{LV}V_L \left[\frac{1}{r_1} + \frac{1}{r_2} \right] \tag{3.12}$$

In general:

$$RT \ln x = \frac{-2\gamma_{LV}V_L \cos\theta}{r_K} \tag{3.13}$$

where r_K is the Kelvin radius.

For nitrogen at liquid nitrogen temperature:

$$
\begin{aligned}
\gamma_{LV} &= 8.85 \times 10^{-3} \text{ N m}^{-2}; \\
V_L &= 34.6 \times 10^{-6} \text{ m}^{-3} \text{ mol}^{-1}; \\
R &= 8.314 \text{ J mol}^{-1} \text{ K}^{-1}; \\
T &= 77 \text{ K}; \\
\theta &= 0°.
\end{aligned}
$$

Substituting in equation (3.13):

$$r_K = \frac{4.15}{\log x} \times 10^{-10} \text{ m} \tag{3.14}$$

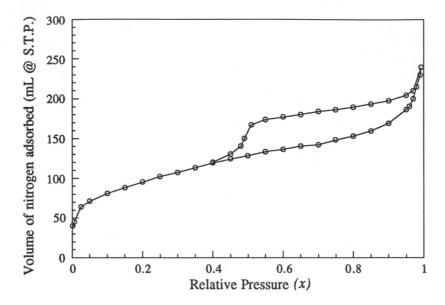

Fig. 3.1 Adsorption of nitrogen on activated clay catalyst. Open circles indicate adsorption; solid circles desorption. Total BET surface, S_w =339 m^3 g^{-1}; monomolecular volume V_m = 78.0 cm^3 at STP.

Pore volue and pore surface distribution may be determined from gas adsorption isotherms. If the amount of gas adsorbed on the external surface is small compared with the amount adsorbed in the pores, the total pore volume is the condensed volume adsorbed at saturation pressure.

3.3 The hysteresis loop

With many adsorbents a hysteresis loop occurs between the adsorption and desorption branches of the isotherm (Figure 3.1). This is due to capillary condensation augmenting multilayer adsorption at the pressures at which hysteresis is present, the radii of curvature being different during adsorption from the radii of curvature during desorption. Since the desorption branch is thermodynamically more stable than the adsorption branch it is usual to use the desorption branch for pore size determination.

Fifteen shape groups of capillaries were analyzed by de Boer [2] from a consideration of five types of hysteresis loop which he designated Type A to Type E (Figure 3.2).

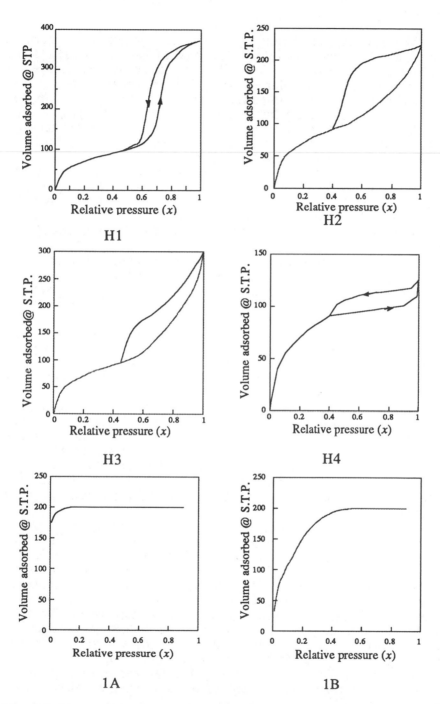

Fig. 3.2 Types of hysteresis loop (IUPAC).

Type A. Both adsorption and desorption branches are steep at intermediate relative pressures. This type of isotherm is designated *Type H1* by x IUPAC [8]. These include tubular capillaries open at both ends; tubular capillaries with slightly wider parts; tubular capillaries of two main dimensions; wide ink bottle pores provided $r_n < r_w < 2r_n$; tubular capillaries with one narrow part; wedge shaped capillaries. IUPAC state that Type H1 can be associated with agglomerates or compacts of uniform spheres in fairly regular array and hence to a narrow distribution of pore sizes.

Type B. The adsorption branch is steep at saturation pressure, the desorption branch at intermediate relative pressure (IUPAC *Type H3*). These include open slit–shaped capillaries with parallel walls; capillaries with very wide bodies and narrow short necks. IUPAC state that this type is observed with aggregates of plate–like particles giving rise to slit–shaped pores

Type C. The adsorption branch is steep, at intermediate pressures the desorption branch is sloping. These are typical of a heterogeneous distribution of pores of some of the following shapes; tapered or double tapered capillaries and wedge formed capillaries with closed sides and open ends.

Type D. The adsorption branch is steep at saturation pressure, the desorption branch is sloping. These occur for a heterogeneous assembly of capillaries with bodies of wide dimension and having a greatly varying range of narrow necks and for wedge shaped capillaries open at both ends.

Type E. The adsorption branch has a sloping character, the desorption branch is steep at intermediate relative pressures (IUPAC *Type H2*). These occur for assemblies of capillaries of one of the shape groups for Type A, when the dimensions responsible for the adsorption branch of the isotherm are heterogeneously distributed and the dimensions responsible for desorption are of equal size. IUPAC state that this provides an over simplified picture and the role of networks must be taken into account.

Sing also described two types of Type 1 isotherms associated with microporous adsorbents having very small external areas. The initial steep region is associated with the filling of pores of molecular dimensions (width < 0.7 nm). The high adsorption affinity, which is a special feature of Type 1A isotherm is due mainly to an enhanced energy of adsorption associated with the strong adsorbent–adsorbate interactions in the very narrow pores. In the case of the Type 1B isotherm, the more gradual approach to the plateau is the result of filling of wider micropores (width 0.1–2 nm) by a secondary process involving co–operative adsorbate–adsorbate interactions

Jovanovic added three more types; for carbon and graphite, charcoal and silica gel and a stepped isotherm on graphitized carbon with krypton [10].

3.4 Theoretical evaluation of hysteresis loops

3.4. 1 Cylindrical pore model

Consider a cylindrical pore open at both ends and of radius r_p (Figure 3.3). During adsorption $r_1 = r_c$, $r_2 = \infty$ where r_c is the core radius:

$$r_c = r_p - t \tag{3.15}$$

where t is the thickness of the condensed vapor in the pores. During desorption the radii are $r_1 = r_2 = r_c$
Inserting in equation (3.12) gives:

$$RT \ln x_A = \frac{\gamma_{LV} V_L \cos \theta}{r_c} \tag{3.16}$$

$$RT \ln x_D = \frac{2 \gamma_{LV} V_L \cos \theta}{r_c} \tag{3.17}$$

Hence

$$x_A^2 = x_D \tag{3.18}$$

where A refers to adsorption and D to desorption. Hence, for a given volume V adsorbed, $x_A > x_D$. For example, the volume adsorbed at $x_A = 0.8$ is the same as the volume desorbed at $x_D = (0.8)^2 = 0.64$. For cylindrical pores closed at one end there is no hysteresis.

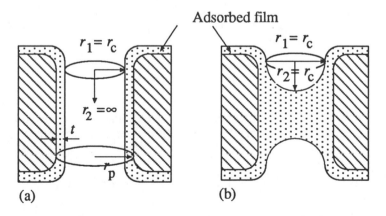

Fig. 3.3 (a) adsorption in and (b) desorption from a cylindrical pore.

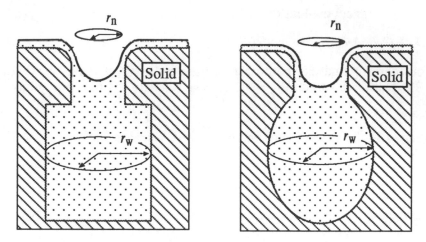

Fig. 3.4 Ink–bottle pores.

3.4.2 Ink bottle pore model:

For tubular capillaries with narrow necks and wide bodies (Figure 3.4) where $r_w \leq 2r_n$, the necks will fill when the Kelvin radius (Equation 3.13) corresponds to $r_n/2$; this will produce a spherical meniscus in the wider parts and increase the pressure there to:

$$P = P_0 \exp\left(\frac{2\gamma_L \cos\theta}{RTr_n}\right)$$
(3.19)

(replacing γ_{LV} with γ for simplification)

which is greater than that required to fill the wider parts:

$$P = P_0 \exp[\frac{\gamma_L \cos\theta}{RTr_w}]$$
(3.20)

The whole capillary will therefore fill at the adsorption pressure for the small capillary. On desorption it empties when the pressure is given by:

$$RT\ln x_D = \frac{2\gamma_L}{r_n}\cos\theta; \qquad \text{hence } x_D = x_A^2$$

With ink bottle pores with wide closed bodies and open short necks, the necks are filled when the Kelvin radius corresponds to $r_n/2$ but it is only at a relative pressure corresponding to $r_w/2$ that the whole capillary is filled. Emptying takes place at a relative pressure corresponding to a Kelvin radius corresponding to r_n. Hence $x_D = x_A^2$ as before.

3.4.3 Parallel plate model:

For parallel plates or open slit–shaped capillaries, a meniscus cannot be formed during adsorption, but during desorption a cylindrical meniscus is already present, hence adsorption is delayed to produce hysteresis. During desorption (Figure 3.5) $r_1 = r$ and $r_2 = \infty$ hence:

$$\frac{2}{r_K} = \left[\frac{1}{r_1} + \frac{1}{r_2}\right] \tag{3.21}$$

$$\frac{2}{r_K} = \frac{1}{0.5(d_p + 2t)} \tag{3.22}$$

The relative pressure at which the hysteresis loop closes depends upon the nature of the adsorbate [11] being around 0.42 for nitrogen. Low pressure hysteresis is associated with inelastic distortion of the solid [12].

3.5 Classification of pores

Dubinin [13,14] classified pores into three categories, the boundary limits being modified slightly by IUPAC:

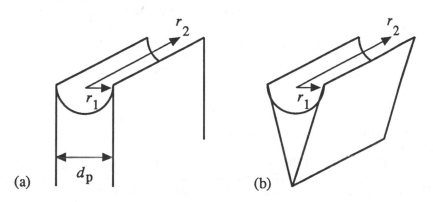

Fig. 3.5 Desorption from (a) parallel plate (slit) and (b) wedge shaped pores.

(1) Macropores having widths in excess of 0.05 μm (50 nm); capillary condensation does not take place in these pores which are essentially avenues of transport to smaller pores.

(2) Mesopores, also known as intermediate or transition pores, having widths between 2 and 50 nm; these mark the limit of applicability of the Kelvin equation.

(3) Micropores having widths not exceeding 2 nm. Since the concept of surface of a solid body is a macroscopic notion, surface area loses its significance when micropores are present, however pore volume remains an applicable concept.

Mikhail and Robens [15] extended this classification to include

(4) ultramicropores of diam eter smaller than the molecular diameter of adsorptives (about 0.6 nm).

3.6 Relationship between the thickness of the adsorbed layer and the relative pressure

From the Kelvin equation, a value of r, say r_1, can be calculated for a given relative pressure x_1 and the volume adsorbed V_1 determined. At a slightly lower relative pressure x_2 the value of r will be r_2 and the volume adsorbed, V_2. If the amount adsorbed on the walls is neglected then V_1-V_2 is equal to the volume of pores in the size range r_1 to r_2 and the cumulative volume of the pores smaller than r_1 will be $V-V_{max}$.

However, as the pressure falls from V_{max} to V_1, nitrogen is desorbed from the core of the smallest pores leaving behind a residual film. On the desorption of the next incremental volume the cores of the next smallest group of pores empties together with some of the nitrogen on the surface exposed by the first desorption.

If allowance is made for the thickness of the residual film t, the relevant radius for the first group of pores would be $r_p = r_1 + t_1$.

Thus the volume adsorbed as the pressure increases is made up of two parts; the volume filling capillary cores and the volume which increases the thickness of the adsorbed layers on the, as yet, unfilled pores. In order to determine the pore size distribution it is therefore necessary to know t.

Oulton [3] and Barret *et al.* [6]. assumed that the thickness of the adsorbed layer remained constant over the whole pressure region. More accurately, t must be related to amount adsorbed. If V_m is the monolayer capacity of a non–porous reference material, the adsorption at any pressure can be converted into film thickness:

$$t = y \frac{V}{V_m} \tag{3.23}$$

where y is the thickness of one layer.

The value of y will depend on the method of stacking successive layers. For nitrogen: if cubical packing is assumed, $y = \sqrt{16.2} = 4.02$ angstroms (0.402 nm). A more open packing [5,16] will give a value of 0.43 nm. For hexagonal close packing [17].

$$y = \frac{MV_s}{N\sigma} \tag{3.24}$$

where

M is the molecular weight of the gas;
V_s is the specific volume;
N is the Avogadro constant;
σ is the area occupied by one molecule.

For nitrogen,

$$y = \frac{28 \times 1.237 \times 10^6}{6.023 \times 10^{23} \times 16.2 \times 10^{20}}$$

y = 0.355 nm

t may be obtained in terms of x by combining the BET equation with equation (3.23) to give:

$$t = \frac{cyx}{(1-x)[1+(c-1)]} \tag{3.25}$$

Since the BET equation predicts too high a value for V in the high pressure region, this equation will also predict too high a value in this region.

Schüll *et al.* [18] showed that for a number of non–porous solids, the ratio between the adsorbed volume V and the volume of the unimolecular layer V_m, if plotted as a function of x, could be represented by a single curve. With the aid of this curve, the thickness of the adsorbed layer could be calculated as a function of x.

Several empirical relationships between film thickness and relative pressure are available. Wheeler [16] suggested that the adsorption on the walls of fine pores is probably greater than on an open surface and proposed the use of Halsey's equation [19]:

$$t^3 \ln(\frac{P}{P_0}) = 5y^3 \tag{3.26}$$

This equation was also used by Dollimore and Heal [20] and Giles *et al.* [21]. Barrett *et al.* [6] used a computation which is, in fact, a tabular integration of Wheeler's equation but they introduce a constant in a manner criticized by Pierce [22]. Their distributions do, however, tend to agree with mercury porosimetry.

Cranston and Inkley [23] derived a curve of t against x from published isotherms on 15 non–porous materials by dividing the volume of nitrogen adsorbed by the BET surface area. They state that their method may be applied either to the adsorption or desorption branch of the isotherm and that the indications were that the desorption branch should be used, a proposal which was at variance with current practice. They assumed cylindrical pores closed at one end but stated that this assumption was unnecessary.

Pierce began with the sample saturated with vapor at P_0 and derived a pore size distribution by considering incremental desorption as the pressure was lowered. He used the cylindrical pore model and applied the Kelvin equation with the assumption that the residual layers were the same as on a non–porous surface at the same pressure. In a later paper [24] he used the Franklin–Halsey–Hill (FHH) equation in the following form:

$$\log(x) = 2.75 \left(\frac{V}{V_m}\right)^{2.99} \tag{3.27}$$

Attempts to improve and simplify earlier models were carried out by Dollimore and Heal [20], who used an open ended non-intersecting cylindrical pore model; Innes [25], who used a parallel plate pore model; and Roberts [26] whose treatment is applicable to both these models. Fifteen papers by de Boer and associates [27–41] made a notable contribution to an understanding of pore systems in catalysis [42]. They found that, for a large number of non–porous inorganic oxides and hydroxides as well as for carbon blacks, the amount of nitrogen adsorbed per unit of surface area is a unique function of relative pressure.

This gives rise to one of the most widely used t-curves which is known as the common t-curve of de Boer [32]. If it is assumed that the adsorbed nitrogen monolayer has the same molar volume as the bulk liquid at the same temperature, then the common isotherm may be represented in the form of a curve representing the thickness of the adsorbed layer as a function of x (Figure 3.6).

Lippens and de Boer [31] published tables of t against x for the construction of this V–t curve from $t = 0.996$ nm at $x = 0.76$ to $t = 0.351$ nm at $x = 0.08$. Broeckhoff [43 cit. 44] extended the curve to $x = 0.92$.

Up to a relative pressure of 0.75 to 0.80 this t curve may be represented by an empirical equation of the Harkins and Jura type [45] ($t < 1.00$ nm):

$$\log x = 0.034 - \frac{0.1399}{t^2} \tag{3.28}$$

For x greater than about 0.4, up to a relative pressure of 0.96, the following empirical relationship holds ($t < 0.55$ nm):

$$\log x = -\frac{0.1611}{t^2} + 0.1682 \exp(-1.137t) \tag{3.29}$$

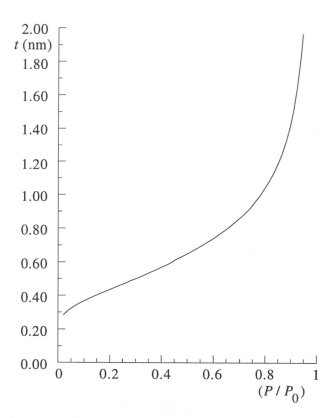

Fig. 3.6 The common t-curve of de Boer.

The two curves are superimposed over the relative pressure range 0.5 to 0.8.

Alternatively, the isotherm may be represented by an equation of the Anderson type [46] with $c = 53$ and $k = 0.76$.

$$\frac{V}{V_m} = \frac{kcx}{(1-kx)(1+(c-1)kx)} \tag{3.30}$$

Broeckhoff and de Boer [47,48] state that the thickness of the adsorbed layer in a cylindrical pore is expected to be different to that on a flat surface at the same pressure and suggested a modified form of the Kelvin equation:

$$RT \ln\left(\frac{P_0}{P}\right) = \frac{\gamma\sigma}{r_p - t} + 2.303RT.F(t) \tag{3.31}$$

where $F(t)$ is given by equations (3.28) or (3.29). This treatment was later extended to ink bottle type pores [49] and applied to cylindrical [50,51] and slit shaped [52] pores.

Dollimore and Heal [53] examined the effect upon the distribution, of changing the method for calculating t, on 36 desorption isotherms and preferred an equation of the same form as equation (3.25) with $y = 0.355$ nm.

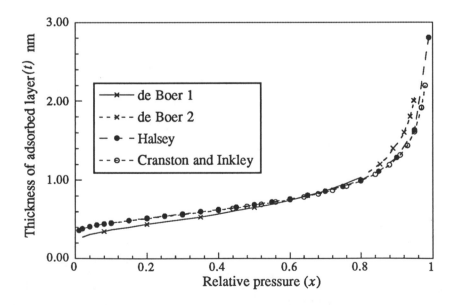

Fig. 3.7 A comparison of some published t curves.

Radjy and Sellevold [54] developed a phenomenological theoretical theory for the *t*-method of pore structure analysis for slit–shaped and cylindrical pores. A comparison [55] of adsorption and desorption methods for pore size distribution, with transmission electron microscopy using closely graded cylindrical pores in alumina, closed at one end, confirmed the superiority of the Broeckhoff-de Boer equations over the Kelvin equation. Lamond [56] found Lippen's *t* silica values unsuitable for carbon black and proposed a new set based on adsorption data for fluffy carbon blacks.

These curves cannot be used for all substances and other *t*-curves are available. When the *t* curve is being used for pore size determination small errors in *t* can be neglected but, for surface area determination, from porosimetry, *t*-curves with small errors are required and these should be common for groups of materials such as halides, metals and graphite [57]. It is also a possibility that the shape of the *t*-curve depends on the BET *c* value [58,59]. A comparison of some of these *t*-curves is presented in Figure 3.7.

The assumption that adsorption on pore walls can be modeled by an isotherm measured on an isolated surface is clearly only valid if the adsorbed films on opposing walls are far apart. If the films are sufficiently close that the molecules on opposing walls interact significantly, the adsorption in the pore will be enhanced relative to that observed on a non–porous solid due to the long–ranged attractive forces between the adsorbed molecules.

3.7 Non linear V–t curves

Surface area may be determined from a plot of volume of gas adsorbed against film thickness [60]. The plot should be a straight line through the origin and the specific surface may be obtained from the slope using equations (2.9) and (3.23) to give:

$$S_t = 1.547 \left(\frac{V}{t} \right) \tag{3.32}$$

with *V* in $cm^3 \, g^{-1}$ (vapor at STP), *t* in nm and S_t in $m^2 \, g^{-1}$.

The V–t method is based on the BET concept but yields additional information. For non-porous solids a graph of *V* against *t* yields a straight line (1). Deviations from the straight line are interpreted as (Figure 3.8):

(2)   decrease in accessible surface area due to blocking of micropores;
(3)   onset of capillary condensation in intermediate (mesopores) pores;

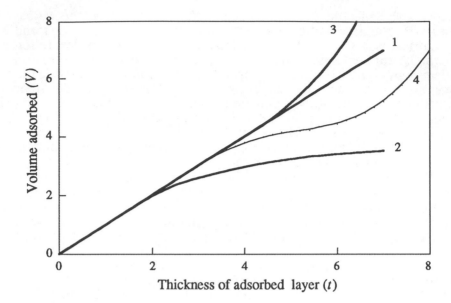

Fig. 3.8 The *V–t* curve.

(4)   Initial decrease in accessible surface due to blocking of micropores followed by onset of capillary condensation in mesopores.

For case (4) a second linear portion gives the remaining surface area, that is, the surface area of the wider capillaries or of the outside area of the granules. The intercept of this second linear portion, on the *V* axis, gives the micropore volume [61].

3.8 The α_S method

The most serious limitation of the *t* method, for surface area determination, is that it is dependent on the BET evaluation of monolayer capacity since *t* is calculated from V/V_m. To avoid this problem, Sing [62–66] replaced *t* with $\alpha_S = V/V_S$ where V_S is the amount adsorbed at a selected relative pressure. In principle α_S can be made equal to unity at any selected point on the isotherm but Sing found it convenient to use a relative pressure of 0.4. Precision is increased by locating α_S in the middle range of the isotherm but higher relative pressures than 0.4 are unsuitable due to the onset of capillary condensation with its associated hysteresis loop. Values of S_S are calculated from the slopes of the linear section of the *V* versus α_S plot by using a normalizing factor obtained from the standard isotherm on a non-porous reference solid of known surface area.

$$S_s = 2.87 \frac{V}{\alpha_s} \qquad (3.33)$$

The micropore volume is obtained from the backward extrapolation of the linear branch of the α_s plot to $\alpha_s = 0$; the intercept on the x–axis gives the effective origin for the monolayer–multilayer adsorption on the external surface. The α_s method has been used for the adsorption of various gases on a range of solids [67]. It has also been used for potential pore size reference materials and the results compared with mercury porosimetry [68]. For nitrogen, a normalizing factor of 2.87 was calculated from the silica TK 600 isotherm [69].

3.9 The n_S–n_R method

This method was proposed by Mikhail and Cadenhead [70], the subscripts standing for (S)ample under test and (R)eference sample, n being the number of adsorbed layers. A plot is constructed of n_S against n_R each value being at the same relative pressure.

This approach eliminates the need to assume a thickness for each adsorbed layer. The authors state that this assumption leads to serious errors if the order of packing and orientation is different in micropores than on plane surfaces.

The thickness t in the t method is an absolute value whereas the number of layers in the n_S–n_R plot is an internal value with each n value being determined from its own particular isotherm. The value n thus includes all the variables, allowing a direct comparison between non–porous and porous materials.

3.10 Relationship between gas volume and condensed liquid volume for nitrogen

The volume of a gas may be reduced to its volume in the condensed phase using the relationship:

$$\Delta V_c = \frac{M \Delta V}{\rho V_L} \qquad (3.34)$$

where ΔV is in cm^3 per gram of adsorbate at standard temperature and pressure (STP). M is the molar weight of the adsorbate, ρ the density of the liquefied gas at its saturated vapor pressure and V_L is the molar volume of the gas at STP. For nitrogen:

$$\Delta V_c = \frac{28}{22400 \times 0.808} \Delta V$$

$$\Delta V_c = 1.547 \times 10^{-3} \Delta V \qquad (3.35)$$

3.11 Pore size distribution determination

The usual way to obtain a pore size distribution is to start with the isotherm at saturated vapor pressure when the pores are completely filled with liquid. A slight lowering of the pressure will result in the desorption of a measurable quantity of vapor. Consider the situation as the relative pressure falls from a value as close to unity as is measurable (i.e. around 0.99) and let this be x_1. As the relative pressure falls from x_1 to x_2, all pores with Kelvin radii greater than r_{K_2} will be emptied except for a residual layer. In the simplest model it assumed that the pores are completely emptied, otherwise it is necessary to assume a residual thickness. The analysis is based on the desorption branch of the isotherm and terminates when the hysteresis loop closes which, theoretically, should occur at a relative pressure greater than 0.4.

3.11.1 Modelless method

The pore shape of very few adsorbents is known and it is unlikely that any one solid will contain pores of only one shape. In the modelless method no pore shape is assumed. The analysis is based on the hysteresis region of the isotherm [71,72]. The method of analysis, strictly speaking, gives the distribution of core volumes and surfaces as functions of the core hydraulic radii (r_h) which is defined as the ratio of the volume to the surface of the cores.

$$r_h = \frac{V_K}{S_K}$$

From equation (3.7):

$$r_h = -\frac{\gamma_{LV} V_L \cos \theta}{RT \ln x} \tag{3.36}$$

Therefore, from equation (3.13), $r_h = (1/2) r_K$.

The cumulative pore size distribution by volume is obtained by plotting V_c against r_h, This may be differentiated graphically to produce the relative pore size distribution by volume or the calculation may be carried out using the tabulated data.

The cumulative pore size distribution by surface is obtained by plotting Sp against r_h where Sp is obtained using equation (3.7) in the form:

$$Sp = -\frac{RT}{\gamma_{LV} V_L \cos \theta} \int \ln x \, dV_c \tag{3.37}$$

Table 3.1 Evaluation of pore size distribution from nitrogen desorption (modelless method)

Relative pressure (x)	Volume desorbed (V_D) cm³ @ STP	Condensed volume (ΔV_{uc}) (uncorr.) cm³ g⁻¹	Relative surface (ΔS_{uc}) (uncorr.) m² g⁻¹	Residual thickness (t) nm	Condensed volume (ΔV_c) (corr.) cm³ g⁻¹	Condensed surface (ΔS_c) (corr.) m² g⁻¹	Mean hydraulic radius (r_h) nm	$\dfrac{\Delta S_{uc}}{\Delta r}$ (uncorr.) m² g⁻¹ nm⁻¹	$\dfrac{\Delta S}{\Delta r}$ (corr.) m² g⁻¹ nm⁻¹	$\dfrac{\Delta V_{uc}}{\Delta r}$ (uncorr.) cm³ g⁻¹ nm⁻¹	$\dfrac{\Delta V}{\Delta r}$ (corr.) cm³ g⁻¹ nm⁻¹
0.99	118.0			6.000			15.31				
0.95	106.5	0.0178	1.16	2.030	0.0178	1.16	5.98				
0.90	98.0	0.0131	2.20	1.430	0.0105	1.75	3.49	0.37	0.30	0.0022	0.0030
0.85	90.5	0.0116	3.32	1.187	0.0104	2.97	2.42	1.87	1.67	0.0065	0.0059
0.80	84.0	0.0101	4.15	1.042	0.0089	3.68	1.83	4.99	4.42	0.0121	0.0125
0.75	78.2	0.0090	4.91	0.938	0.0077	4.22	1.45	10.08	8.66	0.0184	0.0183
0.70	72.8	0.0084	5.76	0.861	0.0072	4.95	1.19	17.93	15.39	0.0260	0.0240
0.65	67.8	0.0077	6.52	0.795	0.0064	5.40	0.99	28.49	23.60	0.0338	0.0313
0.60	62.7	0.0079	7.95	0.739	0.0064	6.49	0.84	46.29	37.76	0.0459	0.0373
0.55	58.0	0.0073	8.63	0.690	0.0057	6.73	0.72	64.31	50.12	0.0542	0.0480
0.50	52.0	0.0093	12.83	0.646	0.0075	10.43	0.63	118.67	96.49	0.0859	0.0524
0.45	45.0	0.0108	17.29	0.606	0.0088	14.09	0.54	193.56	157.75	0.1212	0.0845
0.40	42.0	0.0046	8.52	0.569	0.0023	4.15		14.55	7.09	0.0079	0.0151
Total	76.0	0.1176	83.25		0.0818	66.03					

Table 3.1 (cont.) List of equations used

Core or Kelvin volume	$\Delta V_{uc} = 0.001547 \Delta V_D$
Pore volume (corrected)	$\Delta V_{c,} = \Delta V_{uc} - \Delta t S_c,$
Core surface (uncorrected)	$\Delta S_{uc} = -4939\log(x_{mean})\Delta V_c,$
Hydraulic radius	$r_h = \dfrac{\Delta V_{uc}}{\Delta_{uc}} = \dfrac{\Delta V_c}{\Delta S_c}$
Pore surface (corrected)	$\Delta S_c = -4939 \log(x_{mean})\Delta V_{uc}$
For $0.30 < t < 0.80$:	$t = \sqrt{\dfrac{0.1399}{0.034 - \log(x)}}$
For $0.80 < t < 0.995$:	$\log(x) = \dfrac{0.1611}{t^2} + 0.1682\exp(-1.137t)$

For nitrogen:

$$Sp = -\frac{8.314 \times 78 \times 2.303}{8.72 \times 10^3 \times 34.68 \times 10^6} \int \log x \, dV_c$$

$$Sp = -4.939 \times 10^9 \int \log x \, dV_c \tag{3.38}$$

The limits of integration being the maximum measured relative pressure (circ 0.99) and the relative pressure at which the hysteresis loop closes ($x \geq 0.40$). If the condensed volume, V_c is in cm^3 g^{-1}, Sp is in m^2 g^{-1} with a constant of 4939.

The hydraulic radius is half the Kelvin radius (equation 3.36):

$$r_h = -\frac{2.025}{\log x} \times 10^{10} \text{ m} \tag{3.39}$$

The relative pore size distribution may be determined graphically or from tables. The desorption or adsorption branch of the isotherm can be used but the former is preferred due to thermodynamic stability considerations.

S_p may be evaluated by graphically integrating a plot of log x against condensed volume V_c or by a tabular method [73], an example of which is given in Table 3.1.

It is clear that S_p is not the surface of the walls of the pores but the surface of the cores (the Kelvin or core surface S_K) and V_c is the volume of liquid required to fill the cores (i.e. V_c is the core or Kelvin volume V_K).

Kiselev [74] employed this method successfully for the determination of the total surface area of a number of adsorbents, having only wide pores, and the results were in good agreement with BET surface areas. For narrow pores, core and pore surface differ considerably. In terms of volume distributions, this technique is equivalent to plotting condensed volume desorbed (V_c) against half the Kelvin radius.

The condensed volume desorbed (ΔV_c) is related to the pore surface (ΔS_p) by the following equation:

$$\Delta S_p = \frac{2\Delta V_c}{r_K} \tag{3.40}$$

Agreement with BET surface area should be good if the pores are all wide but considerable differences occur if there are narrow pores present.

In the corrected modelless method [75] a correction is applied for the residual film thickness, i.e. the Kelvin radius is used as $r_K = r_p - t$ making $r_h = 2r_p - 2t$. Since a correction is applied for film thickness the method cannot be considered entirely modelless; however the correction modifies the distribution only slightly (Figures 3.9 and 3.10). Further, using a flat or a curved surface for the residual thickness correction, gives similar results.

Brunauer [76] justifies his method on the grounds that one could use the pore structure analysis in industrial operations. If one has only a single experimental core parameter, the core volume, the derived surface could be in considerable error. In order to confirm a correct pore structure analysis it is necessary that the pore surface be in agreement with the BET surface area. To this Brunauer adds a second criterion, that the cumulative pore volume has to agree with the volume adsorbed at saturation pressure. Earlier investigators could not use this criterion because the Kelvin radius is infinite at saturation pressure, thus the largest pores were left out of consideration. In these pores, though the surface is small, the volume is large. By using the hydraulic radius the whole isotherm can be covered.

Havard and Wilson [77] describe pore measurement on meso–porous silica surface area standard powders. They presented pore size distributions based on the modelless method and the Kelvin equation based on open ended cylinders and spheres with co-ordination numbers of 4, 6 and 8. The isotherm can be used to calibrate BET apparatus over the whole range (samples are available from the British National Physical Laboratory).

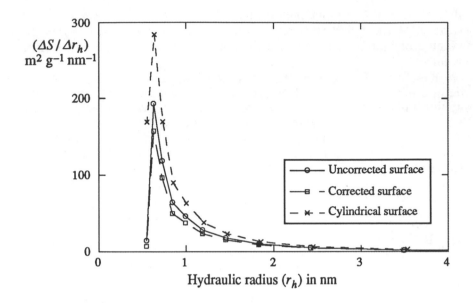

Fig. 3.9 Comparison of pore surface frequency distributions by three methods of calculation.

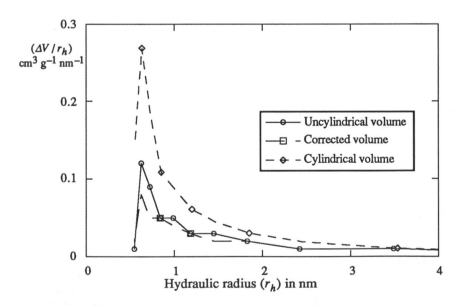

Fig. 3.10 Comparison of pore volume frequency distributions by three methods of calculation.

3.11.2 Cylindrical core model

Using a cylindrical core model open at both ends, for the adsorption isotherm the mutually perpendicular radii are given by: $r_1 = r_c$; $r_2 = \infty$ so that $r_K = 2r_c$, whilst for the desorption isotherm $r_1 = r_2 = r_c$, and $r_K = r_c$.
Thus, from equations (3.16) and (3.17):

$$r_c \ln x_A = -\frac{\mathcal{W}_L}{RT}\cos\theta \qquad\qquad (3.41)$$

$$r_c \ln x_D = -\frac{2\mathcal{W}_L}{RT}\cos\theta \qquad\qquad (3.42)$$

and $x_D = x_A^2$

To obtain the core size distribution either the adsorption or desorption branch of the isotherm can be used but, as before, it is preferable to use the desorption branch.

As the pressure falls from P_{r+1} to P_{r-1} a condensed volume V_{c_r} is desorbed where:

$$V_{c_r} = \pi r^2 L_r(r) \qquad\qquad (3.43)$$

where $L_r(r)$ is the frequency (total length) of cores in the size range $r_{c(r+1)}$ to $r_{c(r-1)}$ centered on r_{c_r}.
The surface of the cores is given by:

$$V_{c_r} = \pi r^2 L_r(r) \qquad\qquad (3.44)$$

Hence, an equation, identical to equation (3.40), is generated:

$$S_{c_r} = \frac{2V_{c_r}}{r_{c_r}} \qquad\qquad (3.45)$$

3.11.3 Cylindrical pore model

This model is a numerical integration method based on the premise that the thickness of the residual layer is the same in the pores as it would be on a plane surface. It was described by Barret, Joyner and Halenda [6] and is known as the BJH method. An application has been presented by Tanev and Vlaev [78].

If allowance is made for the thickness of the adsorbed film, the true pore size distribution is obtained. The full correction is as follows. As the pressure falls from P_0 (in practice a pressure close to P_0 is taken) to P_1, a condensed volume V_{c_1} is desorbed, where V_{c_1} is the core volume of pores with radii greater than r_{P_1} and of average size $\bar{r}_{P_1}$ (Figure 3.11, Step 1). Hence:

$$V_{c_1} = \pi r_{c_1}^2 L_1(r)$$

$$V_{P_1} = \pi r_{P_1}^2 L_1(r)$$

$$S_{P_1} = 2\pi r_{P_1} L_1(r)$$

The volume of the first class of pores is given by:

$$V_{P_1} = \left(\frac{\bar{r}_{P_1}^2}{\bar{r}_{c_1}^2}\right) V_{c_1} \tag{3.46}$$

The surface of the first class of pores is given by:

$$S_{P_1} = \frac{2V_{P_1}}{\bar{r}_{P_1}} \tag{3.47}$$

$\bar{r}_{c1}$ is given by the Kelvin equation:

$$\bar{r}_{c_1} = -\frac{\gamma V_L \cos(\theta)}{RT}\left[\frac{1}{\ln x_0} - \frac{1}{\ln x_1}\right] \tag{3.48}$$

where x_0 is approximately equal to unity and:

$$\bar{r}_{P_1} = \bar{r}_{c_1} + t \tag{3.49}$$

t is obtained from an appropriate t curve. For macropores the choice does not significantly affect the results but the choice becomes more important as the pores get smaller.

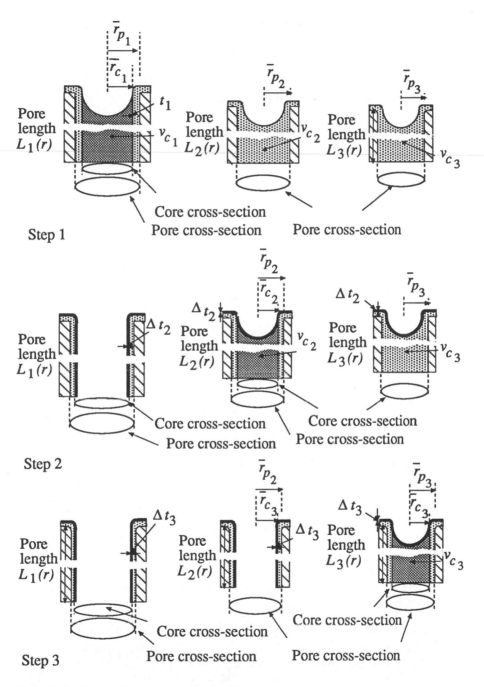

Fig. 3.11 The cylindrical pore model.

As the pressure falls from P_1 to P_2 a condensed volume V_{c2} is desorbed. This consists of two parts (Figure 3.11, Step 2):

1   The core volume of the pores in the size range r_{p_1} to r_{p_2} having a mean size $\bar{r}_{p_2}$ i.e. the Kelvin volume V_K

2   The amount $S_{p_1}\Delta t_2$ desorbed from the exposed surface of the first group of pores where $(\Delta t_2 = t_2 - t_1)$.

Therefore

$$\bar{V}_{P2} = \left(\frac{\bar{r}_{p_2}}{\bar{r}_{c_2}}\right)^2 \left[\bar{V}_{c_1} + S_{p_1}\Delta t_1\right] \tag{3.50}$$

where the core or Kelvin volume is given by:

$$\bar{V}_{K_2} = \left[\bar{V}_{c_2} + \Delta t_2 S_{p_1}\right] \tag{3.51}$$

Also

$$S_{p_2} = \frac{2V_{p_2}}{\bar{r}_{p_2}} \tag{3.52}$$

For the general, or rth desorption step:

$$V_{p_r} = \left(\frac{\bar{r}_{p_r}}{\bar{r}_{c_r}}\right)^2 \left[V_{c_r} - \Delta t_r \sum_{x=0}^{r-1} S_{p_x}\right] \tag{3.53}$$

$$S_{p_r} = \frac{2V_{p_r}}{\bar{r}_{p_r}} \tag{3.54}$$

An application of the procedure is given in Table 3.2, with t derived using equations (3.28) and (3.29), and plots of the pore volume and pore surface distributions are presented in Figures 3.12 and 3.13.

    If the pore size distribution continues below the point at which the hysteresis loop closes, it indicates that the condensation is occurring in pores with shapes not leading to hysteresis such as wedge shaped or conical pores.

The $V_A - t$ curve for this sample is linear up to $t = 0.55$ nm and is then convex to the t–axis, indicating the onset of capillary condensation in mesopores.

The hysteresis loop closes at $x = 0.43$, making total surface area = 191 m^2 g^{-1} which is comparable to the BET value of 200 m^2 g^{-1}. The total pore volume = 0.32 cm^2 g^{-1} which is similar to the volume of intruded mercury (0.34 cm^2 g^{-1} hence the data are in good agreement i.e. the cylindrical model is the correct model to use based on these similarities.

3.11.4 Parallel plate model

During adsorption a meniscus cannot be formed but during desorption a cylindrical meniscus is present. During desorption the Kelvin equation takes the form [31,41] :

$$RT \log x_D = -\frac{2\gamma V_L}{r_K} = -\gamma V_L \left(\frac{1}{d-t} + \frac{1}{\infty}\right) \qquad (3.55)$$

Thus, the plate separation is given by $d = r_K + 2t$.

A similar argument applies to wedge–shaped pores. Assuming that at the highest measured pressure the pores are completely filled, V_{c_1} will be the core volume (V_{K_1}) of the first group of pores of volume V_{p_1} and surface area S_{p_1}.

$$V_{c_1} = V_{K_1}$$

$$V_{K_1} = \frac{S_{p_1}}{2}$$

$$V_{p_1} = \frac{S_{p_1}}{2} \bar{d}_{p_1}$$

From these equations the pore volume and pore surface may be calculated:

$$V_{p_1} = \frac{\bar{d}_{p_1}}{\bar{r}_{K_1}} V_{c_1}$$

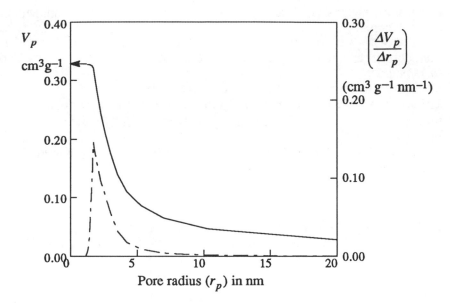

Fig. 3.12 Cumulative pore volume oversize and pore frequency distributions by volume by nitrogen gas porosimetry using a cylindrical pore model.

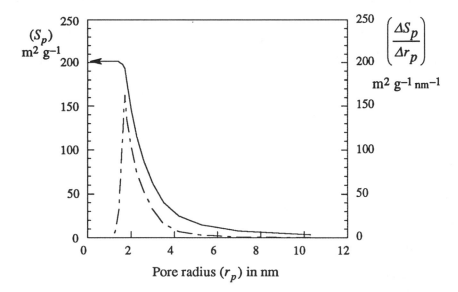

Fig. 3.13 Cumulative pore surface oversize and pore frequency distributions by surface by nitrogen gas porosimetry using a cylindrical pore model.

Table 3.2 Evaluation of pore size distribution from nitrogen desorption isotherm (cylindrical model)

Relative pressure (x)	Volume desorbed (v_D) (cm³ g⁻¹) @STP	Residual thickness (t) nm	Core radius (r_K) nm	Pore radius (r_p) nm	Condensed volume (V_c) (cm³ g⁻¹)	$\bar{r}_K$	$\bar{r}_p$	Cum surface (S_p) (m³ g⁻¹)	Relative pore volume (ΔV_p) (cm³ g⁻¹)	Cum pore volume (V_p) (cm³ g⁻¹)	$\frac{\Delta V_p}{\Delta p}$ (cm³ g⁻¹ nm⁻¹)	$\frac{\Delta S_p}{\Delta p}$ (m³ g⁻¹ nm⁻¹)
0.99	220.0	6.000	93.93	99.93	0.3403					0.028	0.00	0.00
						56.17	60.9	1.00	0.0302			
0.95	203.0	2.030	18.41	20.44	0.3140					0.047	1.92	0.30
						13.68	15.4	3.45	0.0189			
0.90	193.0	1.430	8.96	10.39	0.2986					0.065	5.49	1.30
						7.38	8.69	7.62	0.0181			
0.85	184.0	1.187	5.81	7.00	0.2846					0.086	12.68	4.20
						5.02	6.13	14.62	0.0215			
0.80	174.0	1.042	4.23	5.27	0.2692					0.110	22.40	9.40
						3.76	4.75	24.42	0.0233			
0.75	163.6	0.938	3.28	4.22	0.2531					0.140	41.94	21.70
						2.96	3.86	39.91	0.0299			
0.70	151.0	0.861	2.65	3.51	0.2336					0.176	69.62	42.90
						2.42	3.25	59.30	0.0315			
0.65	138.0	0.795	2.19	2.99	0.2135					0.211	86.60	62.20
						2.02	2.79	84.35	0.0349			
0.60	124.0	0.739	1.85	2.59	0.1918					0.244	104.72	86.20
						1.71	2.43	112.06	0.0336			
0.55	110.5	0.690	1.58	2.27	0.1709					0.277	128.25	119.9
						1.47	2.14	143.61	0.0337			
0.50	97.0	0.646	1.36	2.01	0.1501					0.309	144.50	152.7
						1.27	1.90	177.31	0.0320			
0.45	84.0	0.606	1.18	1.79	0.1299					0.321	147.19	172.1
						1.15	1.75	190.73	0.0117			
0.43	79.0	0.591	1.12	1.71	0.1222							

For $0.30 < t < 0.80$: $t = \sqrt{\dfrac{0.1399}{0.034 - \log(x)}}$ nm

For $0.80 < t < 0.995$: $\log(x) = \dfrac{0.1611}{t^2} + 0.1682\exp(-1.137t)$

$r_c = r_K = -\dfrac{0.410}{\log x}$ nm

$r_p = r_c + t$

$V_c = 0.001547 V_D$

$\Delta v_p = \left(\dfrac{\bar{r}_p}{\bar{r}_K}\right)^2 \left[\Delta V_c - 0.001\Delta t \sum_0^x \Delta S_{px}\right]$

$\Delta S_p = \dfrac{2000\Delta V_p}{\bar{r}_p}$

Table 3.3 Evaluation of pore size distribution from nitrogen desorption isotherm (slit-shaped model)

Relative pressure (x)	Volume desorbed (v_D) (cm³ g⁻¹) @STP	Residual thickness (t) nm	Core radius (r_K) nm	Slit width (d_p) nm	Condensed volume (V_c) (cm³ g⁻¹)	$\bar{r}_K$	$\bar{d}_p$	Cum surface (S_p) (m² g⁻¹)	Relative pore volume (ΔV_p) (cm³ g⁻¹)	Cum pore volume (V_p) (cm³ g⁻¹)	$\dfrac{\Delta V_p}{\Delta d_p}$ (cm³ g⁻¹ nm⁻¹)	$\dfrac{\Delta S_p}{\Delta d_p}$ (m² g⁻¹ nm⁻¹)
0.99	220.0	6.000	93.93	105.93	0.3403							
0.95	203.0	2.030	18.41	22.465	0.3140	56.17	64.20	0.94	0.0301	0.0301	0.000	0.0
0.90	193.0	1.430	8.960	11.820	0.2986	13.68	17.14	3.64	0.0187	0.0487	1.755	0.3
0.85	184.0	1.187	5.809	8.183	0.2846	7.385	10.00	8.46	0.0177	0.0664	4.855	1.3
0.80	174.0	1.042	4.231	6.315	0.2692	5.020	7.249	16.71	0.0206	0.0870	11.009	4.4
0.75	163.6	0.938	3.282	5.158	0.2531	3.756	5.736	28.48	0.0219	0.1089	18.952	10.2
0.70	151.0	0.861	2.647	4.368	0.2336	2.964	4.763	47.23	0.0278	0.1367	35.150	23.7
0.65	138.0	0.795	2.191	3.782	0.2135	2.419	4.075	75.26	0.0339	0.1705	57.858	47.9
0.60	124.0	0.739	1.848	3.327	0.1918	2.020	3.555	105.66	0.0307	0.2012	67.406	66.7
0.55	110.5	0.690	1.579	2.960	0.1709	1.714	3.143	139.27	0.0288	0.2300	78.302	91.5
0.50	97.0	0.646	1.362	2.654	0.1501	1.471	2.807	177.62	0.0282	0.2582	92.245	125.7
0.45	84.0	0.606	1.182	2.395	0.1299	1.272	2.524	218.19	0.0258	0.2839	99.235	156.1
0.43	79.0	0.591	1.119	2.301	0.1222	1.150	2.348	233.83	0.0090	0.2930	96.318	166.6

For $0.30 < t < 0.80$: $t = \sqrt{\dfrac{0.1399}{0.034 - \log(x)}}$ nm

$r_c = r_K = -\dfrac{0.410}{\log x}$ nm

$\bar{r}_K = -\dfrac{0.410}{\log x}$ nm

$d_p = r_K + 2t$

$V_c = 0.001547\,V_D$

For $0.80 < t < 0.995$: $\log(x) = \dfrac{0.1611}{t^2} + 0.1682\exp(-1.137t)$

$\Delta v_p = \left(\dfrac{\bar{d}_p}{\bar{r}_K}\right)^2 \left[\Delta V_c - 0.001\Delta t \sum_0^x \Delta S_{px}\right]$

$\Delta S_p = \dfrac{2000\,\Delta V_p}{\bar{r}_K}$

$$S_{P_1} = \frac{2V_{c_1}}{\bar{r}_{K_1}}$$ (3.56)

When the pressure is lowered from P_2 to P_3 the desorbed volume will consist of two parts, the volume desorbed from the second group of pores plus the volume desorbed from the surface of the first group of pores.

$$V_{c_2} = V_{K_2} + \Delta t_2 S_{P_1}$$ (3.57)

where

$$V_{K_2} = \frac{S_{P_2}}{2} r_{K_2}$$

$$V_{P_2} = \frac{S_{P_2}}{2} d_{P_2}$$

$$S_{P_2} = \frac{2V_{P_2}}{\bar{d}_{P_2}}$$ (3.58)

In general, lowering the pressure from P_{r+1} to P_{r-1}, thus emptying the cores of slits in the size range d_{r+1} to d_{r-1} with average separation d_r is covered by the following equations:

$$V_{P_r} = \frac{\bar{d}_{P_r}}{\bar{r}_{K_r}} \left[V_{c_r} - \Delta t_r \sum_{x=1}^{r-1} S_{P_x} \right]$$ (3.59)

$$S_{P_r} = \frac{2V_{P_r}}{\bar{d}_{P_r}}$$ (3.60)

The experimental data from Table 3.2 has been recalculated in Table 3.3 using this model.

The hysteresis curve closes at $x = 0.43$ making the total surface area $S_p = 234$ m^2 g^{-1} and the total pore volume $V_p = 0.293$ cm^3 g^{-1}. The BET surface area $S_{BET} = 200$ m^2 g^{-1} and the total volume of intruded mercury $V_T = 0.340$ cm^3 g^{-1} hence the calculated data are in poor agreement with other data i.e. the slit model is incorrect.

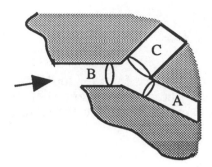

Fig. 3.14 Network of three pores in a porous solid.

3.12 Network theory

There is a growing consensus of scientists who consider the present approach to pore size determination to be unreliable and prefer a different approach. They consider that pore size determination from the desorption branch of the isotherm gives misleading data, particularly if the hysteresis loop is broad.

Network theory describes isotherms in terms of pore connectivity and pore size distribution. At the end of adsorption, when a high relative pressure has been reached and the adsorption isotherm has formed a plateau, all the accessible pores have been filled. On reducing the pressure, liquid will evaporate from the larger open pores but will be prevented from evaporating from equally large pores that are connected to the surface via narrower channels. Desorption more closely reflects the distribution of channels rather than the distribution of pores.

Network theory defines the resulting hysteresis between the adsorption and desorption branches of the isotherm in terms of pore interconnectivity. As the pressure is reduced, a liquid filled cavity cannot convert to the gas phase until at least one of the channels to the outside has evaporated. If the radii of all the channels are less than the equivalent radius of the cavity, the emptying is governed by the largest channel radius, rather than the cavity radius, which will take place at a reduced pressure. Once the liquid in the channel evaporates, the liquid in the cavity can also evaporate and the cavity empties. This is what causes hysteresis – adsorbate in the small channel must evaporate before the adsorbate in the cavity can evaporate. As a result the geometry of the network determines the shape of the desorption branch of the isotherm.

Adsorption provides more accurate pore size information since adsorption takes place through the porous network. This means that even though small cavities fill first they do not block off what is happening in the internal big cavities. Adsorption is a continuous process where adsorbate molecules continue to be transferred to the interior with no resistance.

To illustrate, suppose an increase in external gas pressure causes liquid to condense into a small channel. If the other end is connected to a bigger cavity the liquid would simply evaporate from that end of the channel into the bigger cavity until equilibrium was reached. Adsorption filling is determined by the size of the pore correlated to relative pressure whereas desorption is determined by the branching interconnectivity of the porous network [79].

Seaton [80] describes a method for the determination of pore connectivity based on the use of percolation theory to analyze adsorption isotherms. He illustrated the role of connectivity by the simple example of nitrogen adsorption into three pores (Figure 3.14): of the three pores only pore B is in contact with the exterior of the sample. As the pressure is increased during the adsorption process, nitrogen condenses into the pores in order of increasing pore size i.e. in the sequence A, B, C. In the desorption process, the order in which the liquid nitrogen in the pores becomes thermodynamically unstable with respect to the vapor is phase C, B, A. However the nitrogen in contact with pore C is not in contact with the vapor phase and is unable to vaporize at its condensation pressure. As a result, metastable liquid nitrogen persists in pore C below its condensation pressure, until the liquid in pore B, which is in contact with the vapor phase, vaporizes. The nitrogen in pore A is then in contact with its vapor and is able to vaporize at its condensation pressure. The order of vaporization is thus; B and C together followed by A, with the delay in the vaporization from pore C giving rise to hysteresis. The effect of connectivity on sorption hysteresis may be summarized by observing that nitrogen in a pore of width w is prevented from vaporizing if every path between that pore and the external surface of the solid contains at least one pore of width less than w.

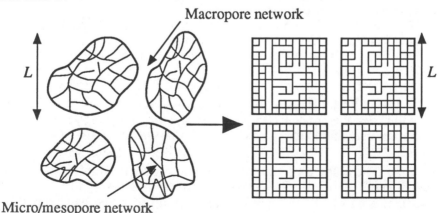

Macropore network

Micro/mesopore network

Fig. 3.15 Mapping of the pore structure of a real solid to a lattice array.

Figure 3.15 is a two–dimensional representation of the structure of a catalyst, which is made up of primary particles containing micro/mesopores which are aggregated together, separated by macropores, to form a pellet. This structure can be mapped as a lattice having the same pore size distribution and the same co–ordination number as the catalyst. Each pore becomes a bond in one of the lattices and each pore junction becomes a node. The size of the pores is assigned to the bonds so that the real structure and the lattice structure have the same pore size distribution, and the mean co–ordination numbers are the same.

In percolation theory, the bonds have two possible states: occupied, denoted by a line in Figure 3.15, or unoccupied. Each bond is occupied with a probability f, which is the same as the fraction of bonds occupied. As f is increased, larger and larger clusters of bonds are formed until a cluster is formed that is large enough to span the lattice. This occurs at a well defined value of f, known as the percolation threshold f_c. This approach is particularly useful in catalyst design and is covered in more detail in various publications referred to by Seaton.

In a later paper [81] the method is improved and generalized by (1) incorporating a more realistic treatment of the desorption process, (2) adapting the method so that it can be used in conjunction with methods based on the Kelvin equation and (3) investigating the effect of pore shape on the results. A further description is contained in [82]. The method turned out to be inadequate for solids containing large mesopores and was later modified to correct for this [83].

3.13 Analysis of micropores; the MP method

In this method, micropores are considered to fill by the growing together of the adsorbed films on opposing pore walls [84]. The thickness of the films is obtained from a t curve derived from measurements of adsorption on non–porous solids.

The micropore isotherm looks very similar to the Type 1 Langmuir isotherm since adsorption is limited to the few layers that can adsorb within the pores and, when these are filled, there is very little external surface remaining [85].

A typical isotherm is shown in Figure 3.16. In this example the BET surface area, using the first four points on the isotherm, is 793 m^2 g^{-1}. Initially the condensed volume adsorbed is $\Delta V_c = S_t \Delta t$ where S_t is the surface of all the pores. The initial slope of the V_c–t curve (Figure 3.17) is $(0.241/0.3043) \times 10^3$ m^2 g^{-1} therefore $S_t = 792$ m^2 g^{-1}. The slope then decreases as one proceeds from $t = 0.3400$ to 0.3678 nm (Table 3.4). The tangent between these two values indicates a surface area of 784 m^2 g^{-1}. Thus a group of pores has been filled with nitrogen and the surface area of these pores is 8 m^2 g^{-1}. The volume of the first group of pores is, therefore, given by:

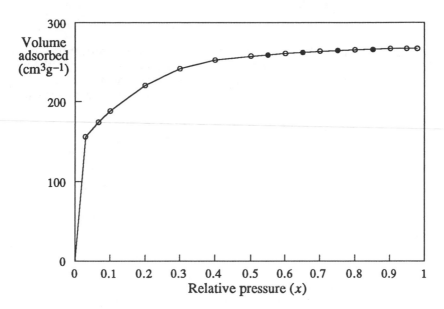

Fig. 3.16 Adsorption–desorption isotherm of nitrogen on silica gel; empty circles, adsorption; black circles, desorption.

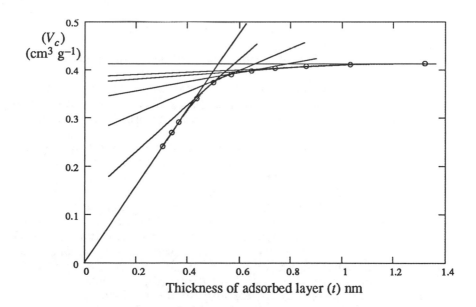

Fig. 3.17 Isotherm of Figure 3.10 converted into a V_C versus t plot.

$$V_p = (S_1 - S_2)\frac{t_1 + t_2}{2} \times 10^4 \qquad\qquad (3.61)$$

One then proceeds in a similar manner to the second pore group with t between 0.3678 and 0.4369 nm. The analysis continues until there is no further decrease in the V_c–t slope which means no further blocking of pores by multilayer adsorption. The pore volume distribution curve is shown in Figure 3.18.

In Table 3.4, column 2 gives the volume adsorbed at relative pressures given in column 1; the thickness of the adsorbed layer *(t)* is derived from equation 3.29; the condensed volume from equation 3.34; the cumulative specific surface from the slope of the v_A versus t curve, equation 3.32, Figure 3.17; the pore volume from equation 3.61 and the hydraulic radius from equation 3.41.

The MP method is based on the use of the appropriate t curve; the choice is far more important in the micropore region than in the mesopore since, in this low pressure region, the heats of adsorption affect the film thickness strongly. Far more important than this, the t values constitute the total pore radius, whereas in the mesopore region they appear only as a correction term.

This approach has been criticized [86] on the grounds that in pores about two molecular diameters wide the influence of opposite walls is significant and once one molecule is adsorbed the pore is effectively reduced in size and fills spontaneously. Moreover, pores three, four and perhaps five diameters wide fill with adsorbate at relative pressures below that at which equivalent numbers of multilayers form on an open surface. This concept of volume filling was introduced earlier by Dubinin and co–workers [87,88].

In reply to this criticism Brunauer examined four silica gels, two containing no micropores and two containing micropores and mesopores. He used nitrogen, oxygen and water as adsorbates and found good agreement between the cumulative pore volume and surface, and the BET surface and the volume adsorbed at the saturated vapor pressure in all cases.

Seaton and co–workers [9] state that because of the unphysical nature of the underlying assumption, this method does not provide a reliable means of determining micropore pore size distributions. Indeed they state that there is no current analysis method that is based on a realistic description of micropore filling, although several semi–empirical approaches have been presented [89–91].

Sing [92] states that there is general agreement that the BET method cannot be used to obtain a reliable assessment of an absorbent exhibiting molecular sieve properties [93,94] (Type 1A isotherms) although the method may be useful for comparison purposes.

Opinions differ on the applicability of the BET analysis to Type 1B isotherms. Indications are that the method can be used to assess the

Table 3.4 Determination of micropore volume distribution from nitrogen adsorption data

x	V_A $(cm^3 g^{-1})$ @STP gas	t (nm)	V_C $(cm^3 g^{-1})$ condensed liquid	S_t $(m^2 g^{-1})$	V_P $(cm^3 g^{-1})$	r_h (nm)	$\overline{r_H}$ (nm)	$\dfrac{dV_P}{dr_h}$ $(cm^3 g^{-1})$
0.033	155.80	0.3043	0.2410	792	0.00	0.138		
0.067	174.10	0.3400	0.2693	792	0.00	0.174	0.157	0.0775
0.100	188.20	0.3678	0.2911	784	0.03	0.205	0.189	0.9728
0.200	219.90	0.4369	0.3402	710	0.12	0.293	0.248	1.0418
0.300	241.30	0.5012	0.3733	515	0.27	0.391	0.340	1.4335
0.400	252.30	0.5691	0.3903	251	0.36	0.514	0.449	0.7637
0.500	257.10	0.6462	0.3977	96	0.38	0.679	0.590	0.1464
0.600	260.80	0.7395	0.4035	61	0.41	0.922	0.788	0.0887
0.700	263.50	0.8606	0.4076	34	0.42	1.320	1.093	0.0395
0.800	265.50	1.0338	0.4107	18	0.43	2.110	1.637	0.0155
0.900	266.90	1.3244	0.4129	7	0.44	4.469	2.897	0.0046
0.950	266.90	1.5767	0.4129	0	0.44	9.180	6.040	0.0000

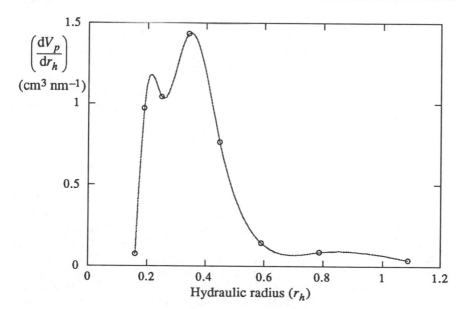

Fig. 3.18 Micropore volume distribution curve.

surface area of the wider micropores (width » 1–2 nm) provided that there are very few narrow pores to distort the isotherm in the monolayer range [95].

The experimental isotherm is the integral of the single pore isotherm multiplied by the pore size distribution [9]. For slit shaped pores this can be written:

$$V_p = \int_{d_{min}}^{d\,max} f(d)\rho(P,d)\mathrm{d}d \qquad (3.62)$$

where V_p is the volume adsorbed at pressure P, d_{min} and $d$max are the widths of the smallest and largest pores present (where the pore width is the distance between the nuclei of the carbon atoms on opposing pore walls), and $\rho(P,d)$ is the molar density of nitrogen at pressure P in a pore of width d. The pore size distribution $f(d)$, is the distribution of pore volumes as a function of pore width.

The properties of fluids in pores has been studied extensively using a statistical mechanical approach know as the mean field density–functional theory [96–98]. In this approach, the fluid properties are calculated directly from the forces acting between the constituent adsorbate–adsorbent molecules.

The predictions of mean–field theory for phase equilibrium in pores are equivalent, in the large pore limit, to the thermodynamic model. However, it provides a more realistic representation of the fluid behavior as the pores become smaller. In particular, it predicts the thickening of the adsorbed layers on the pore walls, and the change from capillary condensation to pore filling at the critical pore size.

In computer simulations, mean field theory [99] provides an accurate description of fluid properties (i.e. density and adsorption isotherm), although not fluid structure near the pore walls. The theory diverges increasingly from simulation data as the pore size is reduced because of the greater influence of short range correlations between fluid molecules, which are neglected in the theory. However, it remains qualitatively correct down to very small pore sizes [100].

The ill–posed nature of equation (3.61) presents a mathematical difficulty in that an infinite number of functions $f(d)$ exist which are consistent with the measured value of V_P [101]. In practice, the large number of experimental data points ensures that the calculated distributions are consistent. Seaton *et al.* [9] assumed that the form of the distribution was bimodal log–normal and used the *t* curve of de Boer and Lippens [27] measured on graphitized carbon blacks. This *t* curve is very similar to *t* curves obtained on metal oxides and silica and is close to the *t* curve of Cranston and Inkley [23], which was constructed from measurements on a wide range of adsorbents. It is recognized that, in order to give an accurate description of adsorption over a range of different solids, a number of slightly different *t* curves are required [1]. To check that the fit was not unduly constrained by the log–normal form of *f(d)*, the bimodal gamma distribution was tested and found to give distributions which were almost identical.

They found that, for carbon black, the smallest measurable pore width was 1.3 nm at a relative pressure of 0.001 and this they extrapolated down to 0.8 nm.

They state that there is scope for improvement to this method by using a more sophisticated model such as the non–local mean field theory [102] or molecular simulation [103].

3.14 Density functional theory

Olivier [104] developed a method, based on the above theory, for looking at all the pores from the smallest to the largest. Traditionally the Kelvin or BJH theory is used for large pores and the *t* plot, Dubinin approach, MP method or the Horvath Kavazoe method for micropores. All but the last are based on mechanistic models; the Horvath Kavazoe is based on a quasi–thermodynamic approach.

Density functional theory describes gas adsorption on the molecular level using statistical thermodynamics. This produces a more highly detailed model of adsorption so that more information can be extracted from the isotherm. Olivier states that using the molecular approach at very low relative pressures produces far better results than classical methods, which can only average bulk properties for a large aggregate of molecules. This is suitable for analyzing mesopore and macropore systems but fails for micropore systems at very low pressures since there are no bulk properties to average.

The theory was applied by Seaton *et al.* [9] who forced the pore size distribution to fit a functional form. Olivier and Conklin [105] extract the pore size distribution from experimental data so that the final results more accurately reflect the structure of the sample.

3.15 Benzene desorption

Benzene adsorption has advantages over nitrogen adsorption in that nitrogen is limited to a maximum relative pressure of around 0.99 due to pressure fluctuations, whereas with benzene the adsorption temperature is close to room temperature so that fluctuations in (P/P_0) are small at pressures near P_0. Further $(\gamma V_L/RT)$ for benzene is 2.2 times the value for nitrogen so that a lower relative pressure is required for a given pore radius, so that benzene isotherms give information down to a smaller size (1.58 cf 25) nm.

For benzene, $\gamma = 28.5$ m Nm^{-1} and $V_L = 89.43 \times 10^{-6}$ m^3 mol^{-1}. The closure point for the hysteresis loop is at $x = 0.22$, [1, p. 155].

Benzene was used by Naono *et al.* [106] for sizing mesopores and macropores of silica of BET surface area 1.57 m^2 g^{-1}. They used the same t curve as Dollimore and Heal [20,53] and a cylindrical model; the desorption arm was used since capillary condensation was slow whereas desorption proceeded more rapidly.

3.16 Other adsorbates

The adsorption of nitrogen, oxygen and water vapor have been compared and found to give good agreement [107,108]. Naono *et al.* [109–111] have also compared water vapor and nitrogen isotherms.

Average pore width *(L)* can be determined from the adsorption of water vapor on active carbons using an equation developed by Tsunoda [112,113]:

$$L = a\log\left(\frac{\Delta S}{S}\right) + d \qquad\qquad (3.62)$$

where a and d are constants, ΔS is the area within the hysteresis loop and S is the area between the adsorption branch of the isotherm and the relative pressure axis.

References

1   Gregg, G.J. and Sing, K.S.W. (1967), *Adsorption, Surface Area and Porosity, 2nd ed.*, Academic Press, N.Y., *104, 142*
2   Boer, J.H. de (1958), *The Structure and Properties of Porous Materials,* Butterworths,*104,108*
3   Oulton, T.D. (1948), *J. Phys. Colloid Chem.*, **52**, 1296, *104, 114*
4   Wheeler, A. (1955), *Catalysis.* , 2, Reinhold, N.Y. p. 118, *104*
5   Schüll, C.G. (1948), *J. Am. Chem. Soc.*, **70**, 1405, *104, 115*
6   Barret, E.P., Joyner, L.G. and Halenda, P.O. (1951), *J. Am. Chem. Soc.*, **73**, 373–380, *104, 114, 116, 127*
7   Kurz, H.P. (1972), *Powder Technol.*, **6**, 167–170, *104*
8   Sing, K.S.W., Everett, D.H., Haul, R.A.W., Moscou, L., Pierotti, R.A., Rouquérol, J. and Siemieniewska, T. (1985), *Pure Appl. Chem.*, **57**(4), 603–619, *105, 110*
9   Seaton, N.A., Walton, J.P.R.B. and Quirke, N. (1989), *Carbon,* **27**(6), 855–861, *105, 140, 142*
10   Jovanovic, D.S. (1969), *Kolloid Z. Z. Polymer*, **235**, 1214, *110*
11   Burgess, C.G.V. and Everett, D.H. (1970), *J. Colloid Interf. Sci.*, **33**, 611–614, *113*
12   Sing, K.S.W. (1973), Special Periodical Report, *Colloid Sci.*, **1**, *Chem. Soc.*, London, p. 25, *113*
13   Dubinin, M.M. (1967), *J. Colloid Interf. Sci.*, **23**, 487–499, *113*
14   Bering, B.P., Dubinin, M.M. and Serpinsky, V.V. (1972), *J. Colloid Interf. Sci.*, **38**(1), 184–194, *113*
15   Mikhail, R.Sh. and Robens, E. (1983), *Microstructure and Thermal Analysis of Surfaces*, Wiley Heyden, *114*

16   Wheeler, A. (1955), *Catalysis.*, *2*, Reinhold, N.Y. p118, *114*
17   Halsey, G.D. (1948), *J. Chem. Phys.*, **16**, 931, *115*
18   Lippens, B.C., Linsen, B.G. and Boer, J.H. de (1964), *J. Catal.*, **3**, 32–37, *115*
19   Schüll, C.G., Elkin, P.B. and Roess, L.C. (1948), *J. Am. Chem. Soc.*, **70**, 1405, *115*
20   Dollimore, D. and Heal, G.R. (1964), *J. Appl. Chem.*, **14**, *116*
21   Giles, C.H. *et al.* (1978), *Proc. Conf. Struct. Porous Solids*, Neuchatel, Switzerland, Swiss Chem. Soc., *116*
22   Pierce, C. (1953), *J. Phys. Chem.*, **57**, 149–152, *116*
23   Cranston, R.W. and Inkley, F.A. (1957), *Advances in Catal.*, ed. A. Farkas, Academic Press, NY, pp. 143–154, *116*, *141*
24   Pierce, C. (1968), *J. Phys. Chem.*, **72**, 3673, *116*
25   Innes, W.B. (1957), *Analyt. Chem.*, **29**,(7), 1069–1073, *116*
26   Roberts, B.F. (1948), *J. Chem. Phys.*, **16**, 931, *116*
27   Lippens, B.C., Linsen, B.G. and Boer, J.H. de (1964), *J. Catal.*, **3**, 32–37, *116*
28   Boer, J.H. de and Lippens, B.C. (1964), *J. Catal.*, **3**, 38–43, *116*
29   Lippens, B.C. and Boer, J.H. de (1964), *J. Catal.*, **3**, 44–49, *116*
30   Boer, J.H. de, Heuvel, A. van den and Linsen, B.G. (1964), *J. Catal.*, **3**, 268–273, *116*
31   Lippens, B.C. and Boer, J.H. de (1965), *J. Catal.*, **4**, 319–323, *116*
32   Boer, J.H.,de, Linsen, B.G. and Osinga, Th.J. (1965), *J. Catal.*, **4**, 643–648, *116*
33   Boer, J.H.de, *et al.* (1965), *J. Catal.*, **4**, 649–653, *116*
34   Boer, J.H.de, *et al.* (1967), *J. Catal.*, **7**, 135–139, *116*
35   Broekhoff, J.C.P. and Boer, J.H. de (1967), *J. Catal.*, **9**, 8–14, *116*
36   Broekhoff, J.C.P. and Boer, J.H. de (1967), *J. Catal.*, **9**, 15–27, *116*
37   Broekhoff, J.C.P. and Boer, J.H. de (1968), *J. Catal.*, **10**, 153–165, *116*
38   Broekhoff, J.C.P. and Boer, J.H. de (1968), *J. Catal.*, **10**, 368–374, *116*
39   Broekhoff, J.C.P. and Boer, J.H. de (1968), *J. Catal.*, **10**, 377–390, *116*
40   Broekhoff, J.C.P. and Boer, J.H. de (1968), *J. Catal.*, **10**, 391–400, *116*
41   Boer, J.H. de, *et al.* (1968), *J. Catal.*, **11**, 46–53, *116*
42   Boer, J.H. de (1959), *The Structure and Properties of Porous Materials*, Butterworths, *116*
43   Broeckhof, J.C.P. (1969), *PhD thesis*, Univ. of Delft, *117*
44   Boer, J.H. de (1970), *Surface Area Determination*, Proc. Conf. Soc. Chem. Ind., Bristol, Butterworths, *117*
45   Whittemore, O.J. Jr and Sipe, J.J. (1974), *Powder Technol.*, **9**, 159–164, *117*
46   Anderson, R.B. (1946), *J. Am. Chem. Soc.*, **68**, 686, *118*

47   Broeckhoff, J.C.P. and Boer, J.H. de (1967), *J. Catal.*, **9**, 8–14, *118*

48   Broeckhoff, J.C.P. and Boer, J.H. de (1967), *J. Catal.*, **9**, 15–27, *118*

49   Broeckhoff, J.C.P. and Boer, J.H. de (1968), *J. Catal.*, **10**, 153–165, *118*

50   Broeckhoff, J.C.P. and Boer, J.H. de (1968), *J. Catal.*, **10**, 368–374, *118*

51   Broekhoff, J.C.P. and Boer, J.H. de (1968), *J. Catal.*, **10**, 377–390, *118*

52   Broeckhoff, J.C.P. and Boer, J.H. de (1968), *J. Catal.*, **9**, 391–400, *118*

53   Dollimore, D. and Heal, G.R. (1970), *J. Colloid Interf. Sci.*, **33**, 508–519, *118, 143*

54   Radjy, F. and Sellevold, E.J. (1972), *J. Colloid Interf. Sci.*, **39**(2), 367–388, *119*

55   Ihm, S.K. and Ruckenstein, E. (1977), J. *Colloid Interf. Sci.*, **6**(1), 146–159, *119*

56   Lamond, T.G. (1976), *J. Colloid Interf. Sci.*, **56**(1), 116–122, *119*

57   Parfitt, G.D., Sing, K.S.W. and Irwin, D. (1975), *J. Colloid Interf. Sci.*, **53**(2), 187–193, *119*

58   Lecloux, A. (1970), *J. Catal.*, **18**(22), *119*

59   Mikhail, R. Sh., Guindy, N.M. and Ali, L.T. (1976), *J. Colloid Interf. Sci.*, **55**(2), 1232–1239, *119*

60   Mikhail, R.S., Brunauer, S.and Bodor, E.E. (1968), *J. Colloid Interf. Sci.*, **26**, 45–53, *119*

61   Sing, K.S.W. (1968), *Chem. Ind.*, 1520, *120*

62   Sing, K.S.W. (1967), *Chem. Ind.*, 829, *120*

63   Sing, K.S.W. (1970), *Proc. Int. Symp. Surface Area Determination*, ed. D.M. Everett and R.H. Ottewill, Butterworths, London, p. 25, *120*

64   Sing, K.S.W. (1971), *J. Oil Color Chem. Assoc.*, **54**, 731, *120*

65   Sing, K.S.W. (1973) Special Periodical Report, *Colloid Science*, 1, Chem. Soc., London, p. 1, *120*

66   Sing, K.S.W. (1976), *Characterization of Powder Surfaces*, ed. G.D. Parfitt and K.S.W. Sing, Academic Press, *120*

67   Bhambhami, M.R. *et al.* (1968), (1972), *J. Colloid Interf. Sci.*, **38**(1), 109–117, *121*

68   Giles, C.H. *et al.* (1978), *Proc. Conf. Structure of Porous Solids*, Neuchatel, Switzerland, Swiss Chem. Soc., *121*

69   Everett, D.H. *et al.* (1974), *Powder Technol.*, **6**, 166–170, *121*

70   Mikhail,, R.Sh. and Cadenhead, D.A. (1975), *J. Colloid Interf. Sci.*, **55**, 462, *121*

71   Brunauer, S., Mikhail, R.S. and Bodor, E.E. (1967), *J. Colloid Interf Sci.*, **24**, 451, *122*

72   Brunauer, S., Mikhail, R.S. and Bodor, E.E. (1967), *J. Colloid Interf Sci.*, **25**, 353–358, *122*

73 Bodor, E.E., Odler, I. and Skalny, J., (1970), *J. Colloid Interf Sci.*, 32(2), 367–370, *130*

74 Kiselev, A.V. (1945), *USP Khim.*,14, 367, *124*

75 Bodor, E.E., Odler, I. and Skalny, J. (1970), *J. Colloid Interf Sci.*, 32(2), 367–370, *124*

76 Brunauer, S (1976), *Pure Appl. Chem.*,48, 401–405, *124*

77 Havard, D.C. and Wilson, R. (1976), *J. Colloid Interf. Sci.*, 57(2), 276–288, *124*

78 Tanev, P.T. and Vlaev, L.T. (1993), *J. Colloid Interf. Sci.*, 160, 110–116, *127*

79 Sing, R. W. (1992), *The Micro Report,* 3(4), publ. Micromeretics, One Micrometretics Dr., Norcross, GA 30093–1877, *136*

80 Seaton, N.A, (1991), *Chem. Eng. Sci.*, 46(8), 1895–1909, *136*

81 Liu, H., Zhang, L. and Seaton, N.A. (1992), *Chem. Eng. Sci.*, 47(17/18), 4393–4404, *137*

82 Liu, H., Zhang, L. and Seaton, N.A. (1993), *J. Colloid Interf. Sci.*, 156, 285–293, *137*

83 Liu, H., Zhang, L. and Seaton, N.A. (1994), *Chem. Eng. Sci.*, 49(11), 1869–1878, *137*

84 Brunauer, S., Mikhail, R.S. and Bodor, E.E. (1968), *J. Colloid Interf. Sci.*, 26, 45, *137*

85 Brunauer, S. (1976), *Pure Appl. Chem.*, 48, 401–405, *137*

86 Marsh, H. and Rand, B. (1970), *J. Colloid Interf. Sci.*, 33(3), 478–479, *139*

87 Dubunin, M.M. (1967), *J. Colloid Interf. Sci.*, 23, 487–499, *139*

88 Bering, B.P., Dubinin, M.M. and Serpinsky, V.V. (1972), *J. Colloid Interf. Sci.*, 24, 451, *139*

89 Carrot, P.J.M., Roberts, R.A. and Sing, KSW. (1988), *Characterization of Porous Solids*, pp. 89–100, ed. K.K. Unger, J.Rouquerol, K.S.W. Sing and H. Kral, Elsevier, Amsterdam, *139*

90 Dubinin, M.M. *Characterization of Porous Solids*, pp. 127–137, ed. K.K. Unger, J.Rouquerol, K.S.W. Sing and H. Kral, Elsevier, Amsterdam, *139*

91 Jaroniec, M. (1988), *Characterization of Porous Solids*, pp. 212–222, ed. K.K. Unger, J.Rouquerol, K.S.W. Sing and H. Kral, Elsevier, Amsterdam, *139*

92 Sing, K.S.W. (1992), *Proc Particle Size Analysis Conf,* Anal. Div. Chem. Div. Royal Soc. Chem., ed. N.G. Stanley–Wood and R. Lines publ. Royal Soc. Chem. pp13–32, *139*

93 Sing, K.S.W. (1989), *Colloids and Surfaces,* 38(11), *139*

94 Sing, K.S.W. (1991), *Fundamentals of Adsorption,* ed. A.B. Mersman and S.E. Scholl, Engineering Foundation, N.Y. p. 6, *139*

95 Carrot, P.J.M., Drummond, F.C., Kenny, M.B., Roberts, R.A. and Sing, K.S.W. (1989), *Colloids and Surfaces,* 37(1), *141*

96 Evans, R. and Tarazona, P. (1984), *Phys. Rev. Lett.*, 52, 557, *141*

97   Evans, R., Marconi, U.M.B. and Tarazona, P. (1986), *J. Chem. Phys.*, **84**, 2376, *141*

98   Evans, R., Marconi, U.M.B. and Tarazona, P. (1986), *J. Chem. Soc. Faraday Trans.*, **2**, 82, 1763, *141*

99   Walton, J.P.R.B. and Quirke, N. (1986), *Chem. Phys. Lett.*, **129**, 382, *141*

100   Peterson, B.K., Gubbins, K.E., Heffelfinger, G.S., Marconi, U.M.B. and Swol, F. van (1988), *J. Chem. Phys.*, **88**, 6487, *141*

101   McEnaney, B, Mays, T.J. and Causton, P.D. (1987), *Langmuir*, **3**, 695, *141*

102   Tarazona, P. (1985), *Phys. Rev. A*, **31**, 2672, *142*

103   Walton, J.P.R.B. and Quirke, N. (1989), *Molec. Simul.* **2**, 361, *142*

104   Olivier, J.P., (1992), *The Micro Report,* **3**(4), publ. Micromeretics, One Micromeretics Dr., Norcross, GA 30093–1877, *142*

105   Olivier, J. P. and Conklin, W. B. (1993), *Langmuir, 142*

106   Naono, H., Hakuman, M. and Nakai, K. (1994), *J. Colloid Interf. Sci.*, **165**, 532–535, *142*

107   Hagymassy, J.Jr. and Brunauer, S. (1970), *J. Colloid Interf. Sci.*, **33**, 317, *143*

108   Hanna, K.M., Odler, S., Brunauer, S., Hagymassy, J.Jr, and Bodor, E.E. (1973), *J. Colloid Interf. Sci.*, **45**, 38, *143*

109   Naono, H. and Hakuman, M. (1991), *Hyomen*, **29**, 362, *143*

110   Naono, H. and Hakuman, M. (1993), *J. Colloid Interf. Sci.*, **158**, 19, *143*

111   Naono, H., Hakuman, M., Jon, M., Sakurai, M. and Nakai, K. (1993), *Ceram. Trans.*, **31**, 203, *143*

112   Tsunoda, R. (1990), *J. Colloid Interf. Sci.*, **137**, 563, *143*

113   Tsunoda, R. (1991), *J. Colloid Interf. Sci.*, **146**(1), 291–293, *143*

4

Pore size determination by mercury porosimetry

4.1 Introduction

Many commercially important processes involve the transport of fluids through porous media and the displacement of one fluid, already in the media, by another. The role played by pore structure is of fundamental importance, and its size distribution determination necessary, in order to obtain an understanding of the processes. The quality of powder compacts is also affected by the void size distribution between the constituent particles. For these reasons mercury porosimetry has long been used as an experimental technique for the characterization of pore and void structure. Although quantitative information is contained in mercury intrusion – extrusion curves it can only be elucidated fully by the use of a theoretical model for pore structure.

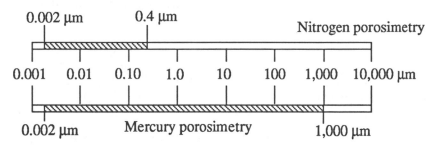

Fig. 4.1 Pore radii ranges covered by gas and mercury porosimetry.

Gas and mercury porosimetry are complementary techniques with the latter covering a much wider size range (Figure 4.1). With judicious choice of the constants in the relevant equations, considerable agreement is found in the overlap regions [1,2].

Mercury porosimetry consists of the gradual intrusion of mercury into an evacuated porous medium at increasingly higher pressures followed by extrusion as the pressure is lowered. The simplest pore model is based upon parallel circular capillaries which empty completely as the pressure is reduced to zero. This model fails to take into account the real nature of most porous media, which consist of a network of interconnecting non-circular pores. Since some of the pores in the body of the particles are not directly accessible to the mercury during the filling cycle, they do not empty during the extrusion cycle and this leads to hysteresis and mercury retention.

The measured pore size distribution is directly affected by pore shape, the relationship between voids and throats (sites and bonds) and the co–operative percolation effects of the porous structure.

4.2 Relationship between pore radii and intrusion pressure

Mercury porosimetry is based on the capillary rise phenomenon whereby an excess pressure is required to cause a non–wetting liquid to climb up a narrow capillary. The pressure difference across the interface is given by the equation of Young and Laplace [3 sic] and its sign is such that the pressure is less in the liquid than in the gas (or vacuum) phase if the contact angle θ is greater than 90° and more if θ is less than 90° (Figure 4.2).

$$\Delta p = \gamma \left(\frac{1}{r_1} + \frac{1}{r_2} \right) \cos \theta \qquad (4.1)$$

where γ is the surface tension of the liquid, r_1 and r_2 are mutually perpendicular radii and θ is the angle of contact between the liquid and the capillary walls (always measured within the liquid).

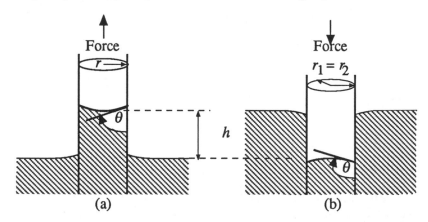

Fig. 4.2 (a) Capillary rise, when liquid wets the wall of the capillary (θ < 90°). (b) Capillary depression, when liquid does not wet the wall of the capillary (90° > θ > 180°).

If the capillary is circular in cross–section, and not too large in radius, the meniscus will be approximately hemispherical. The two radii of curvature are thus equal to each other and to the radius of the capillary. Under these conditions equation (4.1) reduces to the Washburn [4] equation:

$$\Delta p = \frac{2\gamma}{r} \cos \theta \qquad (4.2)$$

where r is the capillary radius.

Mercury porosimetry, in which the amount of mercury forced into a solid is determined as a function of pressure, is based on this equation. If one considers a powder in the evacuated state, $\Delta p = p$, the absolute pressure required to force a non–wetting liquid into a pore of radius r.

For non-wetting liquids (contact angles greater than 90°) the pressure difference is negative and the level of the meniscus in the capillary will be lower than the level in a surrounding reservoir of liquid. In this case Δp is the pressure required to bring the level of the liquid in the capillary up to the level in the surrounding liquid.

For case (a) in Figure 4.2, the pressure above the meniscus is balanced by the hydrostatic pressure drop in a column of liquid of height h and density ρ: an approximate expression for the balancing equation, neglecting the effect of meniscus curvature, may be written:

$$\pi r^2 \rho g h = 2 \pi r \gamma \cos \theta$$

$$\rho g h = \frac{2\gamma \cos \theta}{r} \qquad (4.3)$$

where g is the acceleration due to gravity. This equation also holds for case (b) where $\rho g h$ is the pressure required to bring the two interfaces to the same level.

4.3 Equipment fundamentals

Equipment must possess the facility to evacuate the sample, surround it with mercury and generate sufficiently high pressures to cause the mercury to enter the voids or pores whilst monitoring the amount of mercury intruded.

In almost all porosimeters, the amount of mercury intruded is determined by the fall in the level of the interface between the mercury and the compressing fluid. All porosimeters include certain features in their construction (Figure 4.3).

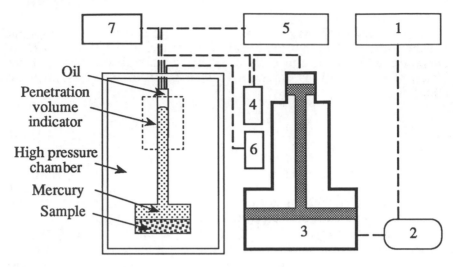

Fig. 4.3 Conceptual representation of a mercury porosimeter. 1, low pressure oil reservoir; 2, pump; 3, pressure multiplier; 4, pressure transducer; 5, high pressure oil reservoir; 6, mercury reservoir; 7, vacuum pump.

These are:

- sample cell;
- vacuum source (and gauge) for degassing the sample;
- source of clean mercury;
- low pressure source and gauge;
- high pressure generator, fluid reservoir and gauge ;
- Ppenetration volume indicator.

The sample is first evacuated and then surrounded with mercury. Air is admitted to the high pressure chamber and the fall in level between the air–mercury interface monitored, to determine the amount of mercury penetrating into the sample, as the air pressure is increased in steps to one atmosphere; the first reading usually being taken at a pressure of 0.5 psia although readings at a pressure of 0.1 psia are possible. This operation is sometimes carried out at a low pressure port. The chamber is then inserted into a high pressure port, the air is evacuated to be replaced by oil, and the pressure is increased to the final pressure of up to 60,000 psia. Commercial instruments work in one of two modes, incremental or continuous. In the former the pressure, or amount of mercury introduced, is increased in steps and the system allowed to stabilize before the next step: in the latter the pressure is increased continuously at a predetermined rate.

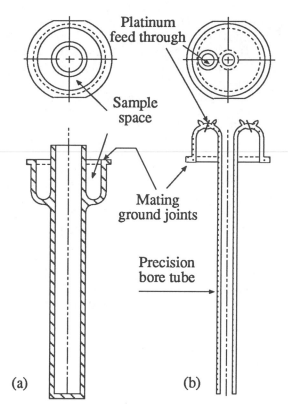

Fig. 4.4 Sample cell for static (incremental) mercury porosimetry.

4.4 Incremental mode

The volume forced into the sample is measured using a penetrometer which is a metal clad, precision bore, glass capillary stem containing the sample (Figure 4.4). Sample cells for incremental mercury porosimetry are available for a wide range of materials including objects as large as one inch cubed, powders, pellets and fabrics.

Vacuum is carefully applied to remove physically adsorbed gases. Degassing times vary depending on the sample and can be greatly reduced if the samples are oven dried before testing. Triple distilled mercury is slowly introduced until it completely covers the sample and fills the sample chamber; any excess is drained off. Air is introduced to raise the pressure to 0.5 psia from which point the analysis begins. The pressure is raised manually or automatically in steps of about 1 psia to atmospheric pressure. As the pressure on the filled penetrometer is increased, mercury intrudes into the sample container and recedes in the capillary. After each increment the pressure is monitored, and may be maintained by the addition of additional pressure until the system comes into equilibrium. The volume of mercury intruded into the sample is determined for each increment. Next, the chamber around the sample cell is filled with hydraulic fluid and the pressure increased in increments to a final value which varies according to the apparatus.

Data are obtained of intruded volume of mercury versus applied pressure and the pressures are converted to pore sizes using the Washburn equation. A full analysis, which may involve fifty or more separate points, can be completed in as little as 30 min for a mesoporous sample but it may take several hours for a microporous sample due to the time required for pressure equilibrium to be reached at each step.

The amount of mercury penetrating into the pores is determined by the fall in level of the interface between the mercury and the hydraulic fluid, correction being made for the compressibility of the mercury and distortion of the interface. This measurement may be carried out:

- using a mechanical follower that maintains contact with the mercury surface as it moves up the dilatometer stem under pressure and relates the distance moved to the volume of mercury intruded;
- by means of the changing resistance of a platinum–iridium wire immersed in the mercury;
- using a capacitance bridge to measure the change in capacitance between the column of mercury in the dilatometer and an external sheath around it.

4.5 Continuous mode

In the continuous (or scanning) mode the pressure is increased continuously from below ambient to some maximum value. In this mode a trained operator can produce up to 12 analyses per hour. The volume of intruded mercury is monitored by means of a capacitance bridge as the quantity of mercury in the stem of a sample cell decreases when filling of the stem occurs (Figure 4.5). A variety of glass sample cells to accommodate a wide range of sample sizes, shapes and porosity are available. From the Washburn equation, at an initial pressure of 0.5 psia, pores and inter-particle voids having radii greater than 213 µm will have filled with mercury. The lower pore radius limit is determined by the maximum pressure achievable in the porosimeter with an absolute limit of 0.0018 µm at a pressure of 60,000 psia with presently available commercial instruments.

Pore size distributions determined in the scanning mode will differ from those obtained in the incremental mode since, in the former mode, the system does not come into equilibrium. However the method is suitable for low porosity samples and quality control purposes.

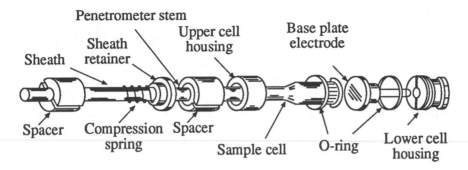

Fig. 4.5 Quantachrome penetrometer assembly for scanning mercury intrusion.

4.6 Discussion

The method was proposed by Washburn in 1921 and the first experimental data were published in 1940 by Henderson, Ridgeway and Ross [5] who used compressed gas to obtain pressures in the range 30 to 900 psia. Ritter and Drake [6,7] extended the range to 10,000 psia, using a compressed oil pumping system, after an accident with a gas system removed the roof of the laboratory. Drake [8] later extended the pressure to 60,000 psia, corresponding to pore diameters greater than 3.5 nm. Further development was carried out by Burdine, Gourney and Reichertz [9] who used dry air at low pressure and nitrogen at high pressure. A simplified apparatus was described by Bucker, Felsenthal and Conley [10]; an instrument for routine determination was designed by Winslow and Shapiro [11] and the first commercial equipment by Guyer *et al.* [12]. Many other modifications to the original equipment have been proposed [13–18] but compressed nitrogen or air was always used to apply pressure to the mercury column. Winslow and Shapiro simplified the operation and improved the safety of mercury porosimetry by using a liquid, isopropyl alcohol, as the hydraulic fluid.

Mercury porosimetry is not applicable where the mercury will come in contact with metals with which it forms amalgams. Glycerin may be used as an alternative [19] since the experimental data determined using glycerin is found to agree with sedimentation balance results [20].

The low pressure region is where inter-particle void filling takes place; pore size distributions frequently have plateau which form a demarcation between voids and pores. To gain maximum information for material having large pores, it is necessary for the initial pressure to be as low as possible and a number of low pressure porosimeters have been developed for this purpose [21]. Leppard and Spencer [22] designed a high pressure apparatus for lump samples having pore radii ranging from 0.35 to 20 nm and Reich [23] for samples in the 0.005 to 50 nm radii range.

Total pore volume may be determined by measuring the density of the material with helium and then with mercury. The difference between the respective specific volumes gives the pore volume.

The measured pore size distribution curves are frequently biased towards the small pore sizes due to the hysteresis effect caused by ink bottle shaped pores with narrow necks accessible to the mercury and wide bodies which are not. Meyer [24] attempted to correct for this using probability theory and this altered the distribution of the large pores considerably. Zgrablich *et al.* [25] studied the relationship between pores and throats (sites and bonds) based on the co–operative percolation effects of a porous network and developed a model to take account of this relationship. The model was tested for agglomerates of spheres, needles, rods and plates. Zhdanov and Fenelonov [26] described the penetration of mercury into pores in terms of percolation theory.

Tsetsekov *et al.* [27] discuss mercury entrapment and breakage in corrugated pores which occurs if the constriction ratio, defined as the diameter of the narrow pore divided by the diameter of an adjacent wide pore, is smaller than some critical value. The amount of mercury entrapment could be predicted over the wide range of from 0 to 85% of intruded volume.

Frevel and Kressley [28] derived expressions for the pressure required for mercury to intrude into a solid composed of a collection of non–uniform spheres. Their treatment defined the pressure required to 'break through', in terms of the largest accessible opening, to the interior of the solid and then related the size of the openings to the size of the spheres. Their model was restricted to a maximum porosity of 39.5%, which was later extended to 47.6% by Meyer and Stowe [29]. This model was later verified by Svata and Zabransky [30] who compared their results with microscopy and sedimentation. Rootare and Craig [31] suggested that both pressurization and de–pressurization curves were required, where the former give size distribution based on the sizes of the 'necks' and the latter the volumes of the pores or voids behind the 'necks'.

Agglomerated particles of different shapes exhibit substantial differences with mercury porosimetry. Specifically, the ratio of intrusion to retraction as a function of void fraction filled varies in different manners. It is suggested that this effect may be used to infer the shape of sub-particles in compacts [32].

The Reverberi method [33] utilizes the difference between the ascending and descending branches of the hysteresis curve for evaluating the broad and narrow parts of the pores independently of each other. The ascending branch is measured in the usual way and the descending branch is measured in steps until the minimum pressure is reached. Svata [34] tested this method using powder metallurgical compacts and found that it was applicable if the powder was first coated with stearic acid to eliminate the interaction between the mercury and the metal. The compacts were coated by immersion in a 12% solution of stearic acid in chloroform.

Adkins *et al.* [35] examined Mo–Al_2O_3 catalysts by SEM, x–ray diffraction and mercury porosimetry and found the porosimetry data consistent with structure as the Mo increased from a single surface layer up to exceeding monolayer coverage. BET surface area remained constant during these changes but the mercury porosimetry area decreased with Mo loading, with the Broekhoff de Boer gas adsorption model giving best agreement with mercury porosimetry. Johnson [36] examined CoO–MoO_3–Al_2O_3 catalysts by nitrogen adsorption and mercury intrusion and considered the latter to be unreliable since it indicated an increase in pore surface with increasing MoO_3, whereas the nitrogen data remained constant. This he attributed to a changing contact angle as the Al_2O_3 became coated with MoO_3.

An examination of the sintering process by mercury porosimetry reveals that, under some circumstances, although the total pore volume decreases with sintering time, the average pore size increases [37]. The technique has been used to investigate the micro structure of tablets [38] as well as a wide range of other materials [39–42].

Hearn and Hooton [43] found that a reduction in sample mass resulted in a decrease in total intruded porosity and that, although the rate of pressure application had no effect on the total intruded volume, it did shift the pore size distribution curves.

4.7 Limitations of mercury porosimetry

The limitations of the technique are:

- The pores are not usually circular in shape and so the results can only be comparative.
- The presence of ink–bottle pores or other shapes with constricted necks opening into large void volumes. All the volume will be assigned to the neck diameter so that the capillaries will be classified at too small a radius. This also leads to hysteresis and mercury retention in the pores [44].
- The assumption of a constant surface tension for mercury.
- The largest source of error in calculating pore diameter is the assumption of a constant value for the angle of contact of mercury. The receding angle of contact may well differ from the advancing angle, i.e. the angle of contact between the mercury and the wall as it advances into the pores under increasing pressure may differ from the contact angle as it recedes under reducing pressure. The contact angle may also be affected by surface roughness.

- The effect of compressibility of mercury, sample container, sample and residual air with increasing pressure.
- Breakdown of the porous structure by the high mercury pressure during intrusion.
- Time effects. Dynamic porosimeters, in which the pressure changes continuously, generate different pore size distributions than static porosimeters, where the pressure increases in steps, with time for the system to come to equilibrium between steps.
- Difficulty in degassing of solids with fine pore structure where the existence of an adsorbed layer reduces both the effective pore diameter and pore radius.

Despite these limitations mercury porosimetry has proved to be a useful tool with which to investigate the internal structure of solids. It should not be considered an absolute method and care should be used in interpreting data.

Table 4.1 Real frequency and measured frequency of an interconnected pore system

Pressure (psia)	Pore radius (µm)	Frequency (real)	Frequency (measured)
100	1.08	1	0
200	0.54	7	1
300	0.36	18	2
400	0.27	28	43
500	0.216	19	26
600	0.180	11	4
700	0.154	5	5
800	0.135	1	1

Direction of flow

6	3	4	2	5	7	4	3	4
4	6	3	4	7	5	2	4	5
5	3	6	4	1	4	7	3	4
4	8	4	3	4	7	3	4	5
5	3	4	4	6	3	2	5	6
5	4	2	5	4	5	4	5	3
4	6	3	5	5	4	3	4	3
3	6	4	4	2	4	5	5	5
5	2	7	4	6	3	4	2	6
5	6	3	5	3	5	4	6	4

Fig. 4.6 Diagrammatic representation of interconnected pores.

4.8 Effect of interconnecting pores

Assume an interconnected network, as in Figure 4.6, where the numbers are the breakthrough pressures in psia /100 from Table 4.1, arranged in a random manner. Let the mercury intrude from the top of the assembly so that at a pressure of 100psia no mercury can penetrate the system; at a pressure of 200psia the mercury can penetrate one pore in the top line; at 300psia the mercury can penetrate two pores in the top line, and at 400psia four pores in line one, four in line two, five in lines three, four and five, four in line six, six in line seven (plus an interconnected pore in line six), five in line eight, four in line nine and one in line ten making a total of 43 pores. The true and measured distributions are presented in Table 4.1 and it can be seen that the effect of interconnected pores is to skew the pore distribution towards the high pressures, i.e. smaller radii.

4.9 Hysteresis

After compression and mercury intrusion; during the extrusion cycle, some of the mercury is trapped in the porous structure. This is due to:

1. *Network effects.* The voids between particles and the pores within them are interconnecting and narrow inlets lead to wide voids. Because of this the experimental data always overweights the narrower pores. During extrusion, mercury is trapped in the wide voids causing hysteresis (see Figures. 4.7 and 4.8).

    Several experimental studies using transparent models indicate that intrusion is governed by the capillary resistance of the pore necks whereas mercury retention during extrusion is controlled by the capillary resistance of pore bodies [45–47].

    Snap–off of threads of mercury, which results in mercury retention, has also been studied extensively [48,49].
2. *Ink bottle pores.* These are pores with narrow necks and wide bodies and they cause the same effect described above. Meyer [50] attempted to correct raw data using probability theory and this considerably altered the measured distribution.
3. *Pore potential.* In gas adsorption, it is established that overlapping potential from the walls of small pores induces condensation at pressures well below the saturated equilibrium vapor pressure. Similarly, mercury once forced into a pore will interact with the pore walls and be trapped in a potential well of energy U. During depressurization, a cylindrical pore of radius r will require an amount of energy E to separate the column such that:

$$E = 2\pi r^2 \gamma \tag{4.4}$$

If $U \geq E$, the column will break, mercury will be trapped and this will cause the extrusion curve to intercept the volume axis above the zero.

During pressurization the mercury column is under compression as it intrudes, whilst under depressurization the column is under tension, due to pore potential, and can break if the pore potential is high and the pore radius very small.

In the case of hysteresis, it is necessary that the pore potential causes mercury to extrude from a pore at a lower pressure than it intrudes. Any shape of pore can result in hysteresis and cause mercury entrapment in the first intrusion.

4. *Surface roughness*. This can cause the mercury to 'slip–stick' so that the mercury thread is broken.
5. *Mercury contamination of the surface*. This has the same effect as (4) as above.

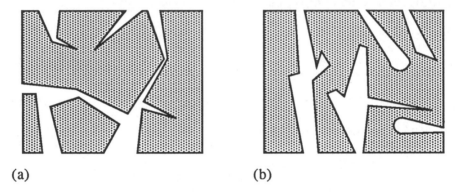

(a) (b)

Fig. 4.7 Channel system (a) that can be drained and (b) that cannot be drained during mercury porosimetry.

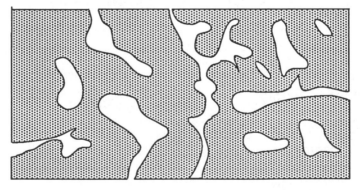

Fig. 4.8 Model of a real pore and channel system [51].

Liabastre and Orr [52] examined graded series of controlled pore glasses and Nuclepore membranes by electron microscope and mercury porosimetry in order to determine compressibilities and explore the reason for hysteresis.

Kloubek [53] considers the concept of pore dimension to be erroneous because of the above errors and recommended that the results be presented using the actual values of *p* instead of calculated radii.   He suggested that the dependence of net re–intrusion and retention volumes on mercury pressure should be evaluated [54].   In this way pores can be separated into two groups, one in which mercury is retained reversibly and the other where retention is irreversible. This method of mercury porosimetry evaluation offers a valuable contribution to understanding porous structures and their properties.

4.10 Contact angle for mercury

As an average for many materials, Ritter and Drake [6,7] used 140° for the mercury contact angle. Winslow and Shapiro [11] found 130° to be a valid value for mercury intruding into a nickel block with 70 drilled holes of 560 µm diameter. A value of 125° was found for intrusion into Portland cement [55]. Juhola and Wiig [56] measured the pressure required to force mercury into a hole 381 µm in diameter and also into a fine calibrated capillary and, in both cases, found the angle to be 140°. However, the values of θ determined on external surfaces and artificially bored holes [57] may differ from the values found in natural pores.

In an extreme case, by choosing a value of 130° or 140° for the contact angle when the real value is 160° or 110°, Rootare [44] demonstrated that the measured pore diameter could be in error by more than 50% .

Using the phenomenon called local hysteresis, values of $\theta = 162°$ and 130° have been obtained for active carbon [55] and a copolymer respectively [58]. Guo *et al.* [59] also found that changing θ from 110° to 180° caused a significant change in the mode of the pore size distribution.

It has also been suggested that contact angle is pore size dependent [60], a value of 180° being applicable for macropores and mesopores with a reduction for smaller pores [61].

The angle of contact during intrusion may be different to the angle during retraction. Mercury contact angles are usually measured by the sessile drop method on flat surfaces although instruments called anglometers are available for measuring the angle in a capillary.

Intrusion and extrusion calculations are conventionally made using the same angle. Lowell and Shields [62,63] using values of θ =170° for intrusion and θ = 107° for extrusion removed all traces of hysteresis apart from that caused by mercury retention.

The use of an incorrect contact angle does not matter when porous materials of the same type are being compared. However, if an exact measurement of pore openings is required it is necessary to either measure the contact angle directly [3, p. 268] or look at the pores under a microscope to establish the relationship between the actual pore size openings to that measured by mercury intrusion.

4.10.1 Effect of contact angle

It is not common practice to measure the contact angle and, indeed, there is some evidence that its value is affected by surface roughness. The hysteresis encountered in mercury porosimetry may be completely removed by the use of different contact angles for intrusion and extrusion. On this basis the use of hysteresis to predict pore shape may lead to erroneous and misleading data.

The method of contact angle measurement adopted by Heertjes and Kossen [64] involves compressing a tablet of the powder and this is open to the objection that the compression process may modify the powder and hence its wetting characteristics. A review of methods for measuring contact angles was presented by Brockel and Löffler [65] and a new method introduced in which the powder is coated on to a carrier plate and the contact angle at the interface between water and cyclohexene measured. Extrapolating data at different porosities gave reproducible results at zero porosity.

4.11 Surface tension of mercury

According to Rootare [42,66] the best of the published values for the surface tension of mercury is given by Roberts [67] who found a value of 0.485 N m^{-1} at 25°C with a temperature coefficient of -0.00021 N m^{-1} °C^{-1}. He attributes the wide variation in published results to the use of contaminated mercury.

Using the above, at 50°C $\gamma = 0.480$ N m^{-1}. For $\Delta p = 200$ MPa and θ =140° [equation (4.2)] the values of γ at 25°C and 50°C yield calculated radii of 3.72 nm and 3.68 nm respectively. The effects of temperature on surface tension values are found to have a minimal effect on measured pore structure and pore size distribution [59].

A higher error in γ values is caused by neglecting the curvature of the meniscus in the pores. The following correction has been suggested [68]:

$$\gamma_{corr} = \gamma - 5.33 \times 10^4 \left(\frac{\Delta p}{2} \right) \tag{4.3}$$

with γ in N m^{-1} and Δp in MPa. For $\Delta p = 200$ MPa the correction term gives an error of 12% [$\gamma_{corr} = (0.485 - 0.053)$ N m^{-1}].

4.12 Corrections for compressibility

In order to calculate the true volume intrusion of mercury into the pores of a sample, a correction must be made to account for the compression of mercury, sample cell and sample [69]. The usual procedure is to carry out a blank experiment in the absence of a sample or with a non–porous sample [70]. During the course of calibration measurements on non–porous nylon it was found that a normal blank correction procedure led to erroneous mercury penetration volumes [71]. In particular it was found that the shape of the intrusion curve varied with the size of the sample.

The rigorous approach to the correction factors, detailed below, resulted in equations which involved the mercury fill volume in the sample cell, the compressibilities of the mercury, glass and sample and the volume of the sample. Figure 4.9 shows the uncorrected intrusion curve for nylon samples together with partial and full corrections.

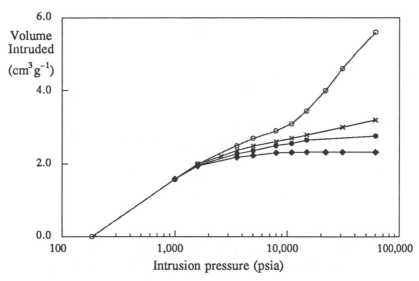

Fig. 4.9 Pressurizing curve for nylon. ^{o}uncorrected curve, ^{x}corrected by a blank run [Equation (4.6)], •corrected for compressibility of mercury and glass [Equation (4.7)], ◆corrected for compressibility of mercury, glass and sample [Equation (4.9)].

On filling the cell with mercury, a near–complete fill is achieved, $V_1-\Delta V_1$. Using a mercury follower (i.e. an automatic calibrated screw probe moves down to the mercury surface where it short circuits and stops a digital counter measuring the probe movement) the small unfilled volume ΔV_1 is given by:

$$HF = \Delta V_1 \tag{4.5}$$

where H is the probe count and F is the volume corresponding to each probe count, known as the cell factor.

The true blank correction $J = h-H$ where h is the probe count at pressure p is given by:

$$JF = p\left[-\Delta V_1\left(\psi_{Hg} - \frac{2}{3}\psi_g\right) + V_1\left(\psi_{Hg} - \psi_g\right)\right] \tag{4.6}$$

$$\left[\psi_{Hg} = 3.55 \times 10^{15} \text{ m}^2 \text{ N}^{-1}, \quad \psi_g = 2.60 \times 10^{15} \text{ m}^2 \text{ N}^{-1}\right]$$

where ψ is the compressibility; the suffixes refer to mercury and glass.

The first term accounts for the incompleteness of fill and the change in the cell factor with pressure and the second term the net compression.

If a non-porous material of mass m, density ρ, and volume V at zero pressure and V_N at pressure p, is incorporated into the cell, the intrusion at pressure p is given by:

$$\frac{J}{m}F = p\left[-\Delta V_2(\psi_{Hg} - \frac{2}{3}\psi_g) + V_1(\psi_{Hg} - \psi_g) + V(\psi_N - \psi_g)\right] \tag{4.7}$$

where J is the intruded count minus the count at zero pressure, ΔV_2 is the initial unfilled volume and ψ_N the compressibility of the solid.

Lee and Maskell [71] used two samples of nylon, one of volume V' and the other of volume V'' to give readings J' and J'' with unfilled volumes $\Delta V_1'$ and $\Delta V_2''$. Equation (4.6) may then be written:

$$\frac{J'}{m'}F = p\left[-\Delta V_1'\left(\psi_{Hg} - \frac{2}{3}\psi_g\right) + V_1\left(\psi_{Hg} - \psi_g\right) + V_N'\left(\psi_N - \psi_g\right)\right] \tag{4.8a}$$

and

$$\frac{J''}{m''}F = p\left[-\Delta V_1''\left(\psi_{Hg} - \frac{2}{3}\psi_g\right) + V_1\left(\psi_{Hg} - \psi_g\right) + V_N''\left(\psi_N - \psi_g\right)\right] \tag{4.8b}$$

Rearranging and assuming

$$\frac{V'}{m'}\left(\psi_N - \psi_{Hg}\right) = \frac{V''}{m''}\left(\psi_N - \psi_{Hg}\right)$$

leads to

$$\left(\frac{J'}{m'} - \frac{J''}{m''}\right)F = p\left[\left(\frac{\Delta V_1'}{m'} - \frac{\Delta V_2''}{m''}\right)\left(\psi_{Hg} - \frac{2}{3}\psi_g\right) + \left(\frac{1}{m'} - \frac{1}{m''}\right)V_1\left(\psi_{Hg} - \psi_g\right)\right]$$

(4.9)

V_1 was taken as 30 cm^3: a fit with experimental J' values was obtained using empirical values of J'' and p.

4.13 Structural damage

Problems can arise due to damage to pores under pressure and fracture can occur to open up previously blind pores [69,72].

Changes in porous structure have been detected by a comparison of the first intrusion curve with one obtained with the same sample after the mercury retained in the pores had been removed by distillation [73–75] Koubek [73] showed that structural damage could be detected by careful examination of intrusion, extrusion and re–intrusion curves.

4.14 Delayed intrusion

Commercial mercury porosimeters operate either in the static or scanning mode. In the static mode, the pressure is increased in steps, equilibrium being attained at each stage; in the scanning mode the pressure is increased continuously. With some samples the rate of mercury intrusion is sufficiently slow that a run can take up to 2 h although more commonly a run is completed in a few minutes. Winslow and Diamond [76] attributed the long times to reach equilibrium to the inability of liquid mercury to rapidly penetrate the narrower regions of a matrix of interconnecting pores. Lowell and Shields [77] proposed an alternative explanation. If a constriction of less than 0.18 μm radius exists within a pore, liquid mercury penetrates to the constriction and no further at 60,000 psia. However the mercury meniscus at the constriction will exhibit a radius of 0.18 μm with an equilibrium relative vapor pressure of 11.49. Mercury vapor transfer continues under the driving force of condensation, leading to a slow approach to equilibrium.

4.15 Theory for volume distribution determination.

Experimental data are obtained in the form of p against V. p is converted to pore radius using equation (4.2): Assuming a surface tension for mercury of 0.480 N m^{-1} and a contact angle of 140° the pore radius in nanometers is given by:

$$r = \frac{0.735}{p} \qquad (4.10)$$

If p is in psia the constant becomes 107 with r in microns.

If the pore size range is narrow it is possible to plot the cumulative pore size distribution by volume on linear paper and to differentiate the curve to obtain the relative pore size distribution by volume. Graphical differentiating smoothes out experimental errors; alternatively, tabular differentiating can be used with a curve smoothing correction.

For a wide distribution it is preferable to plot the data on log–linear paper with V on the linear (y) axis and pore size on the logarithmic axis. The curve of p against V is called the pressurizing curve and from it the various distributions can be found.

If the total volume of pores having radii between r and $r + \delta r$ is dV, the relative pore frequency by volume is defined by:

$$D_3(r) = \frac{dV}{dr} \qquad (4.11)$$

From equation (4.2):

$$pdr + rdp = 0 \qquad (4.12)$$

Combining these two equations gives:

$$D_3(r) = \frac{p}{r}\frac{dV}{dp} \qquad (4.13)$$

For a wide distribution of pore sizes the alternative form of equation (4.13) is preferred:

$$D_3(r) = \frac{1}{r}\frac{dV}{d\ln p} \qquad (4.14)$$

4.16 Theory for surface distribution determination

4.16.1 Cylindrical pore model

Using a cylindrical pore model:

$$\Delta V = \pi r^2 \Delta L \tag{4.15}$$

$$\Delta S = 2\pi r \Delta L \tag{4.16}$$

Defining the relative surface distribution as:

$$D_2(r) = \frac{dS}{dr} \tag{4.17}$$

$$D_2(r) = \frac{dS}{dV}\frac{dV}{dr} \tag{4.18}$$

$$D_2(r) = \frac{2}{r}\frac{dV}{dr} \tag{4.19}$$

By combining with equation (4.12), equation (4.19) may be written as:

$$D_2(r) = \frac{2}{r^2}\frac{dV}{d\ln p} \tag{4.20}$$

4.16.2 Modelless method

Rootare and Penslow [66] obtained surface areas from mercury intrusion data using no assumption of any specific pore geometry. The problem was approached from the point of view that work is required to force mercury into the pores, the work, dW, required to immerse an area δS of powder being:

$$d W = p dV = -\gamma \cos \theta dS \tag{4.21}$$

Therefore

$$\frac{dS}{dr} = -\frac{p}{\gamma \cos \theta}\frac{dV}{dr} \tag{4.22}$$

where $p = \Delta p$

Integrating over the whole range of pressures:

$$S = - \int_0^V \frac{pdV}{\gamma_{LV} \cos \theta} \tag{4.23}$$

For $\gamma = 0.480$ N m^{-2}; $\theta = 140°$; m, the sample mass in grams, p in psia, and S in m^{2}g^{-1}

$$S = - \frac{0.0223}{m} \int_0^V pdV \tag{4.24}$$

This is identical with the integral of equation (4.19) with substitution of r from equation (4.2).

Alternatively, if S is known from BET gas adsorption, this equation may be used to determine $\cos\theta$.

$$\cos(\theta) = - \frac{0.01436}{mS_w} \int_0^V pdV \tag{4.25}$$

Using this equation Rootare [42] found $121°< \theta <160°$ for a range of powders.

4.17 Theory for length distribution determination

Defining the length distribution as $D_1(r)$:

$$D_1(r) = \frac{dL}{dr} \tag{4.26}$$

Combining with equation (4.15) gives:

$$D_1(r) = \frac{1}{\pi r^2} \frac{dV}{dr} \tag{4.27}$$

but $Pr = $ constant hence $pdr = -rdp$ and:

$$D_1(r) = \frac{1}{\pi r^3} \frac{dV}{d\ln p} \tag{4.28}$$

Table 4.2 Evaluation of pore size distribution by mercury porosimetry
(narrow size range)

Pressure (p) psia	Volume intruded (V) cm³ g⁻¹	Pore radius (r) µm	Mean pressure $(\bar{p})$ psia	Mean radius $(\bar{r})$ µm	Specific surface (S) m² g⁻¹	$\dfrac{\Delta S}{\Delta r}$ m² g⁻¹ µm⁻¹	$\dfrac{\Delta V}{\Delta r}$ cm³ g⁻¹ µm⁻¹
			400	0.270		0	0.00
400	0.000	0.270	600	0.180	0.19	1	0.13
800	0.017	0.135	1,000	0.108	0.83	14	0.76
1,200	0.051	0.090	1,400	0.077	2.27	64	2.44
1,600	0.106	0.068	1,800	0.060	5.16	214	6.37
2,000	0.192	0.054	2,200	0.049	9.93	530	12.89
2,400	0.308	0.045	2,600	0.042	14.11	650	13.38
2,800	0.394	0.039	3,000	0.036	16.47	489	8.71
3,200	0.436	0.034	3,400	0.032	18.25	475	7.47
3,600	0.464	0.030	3,800	0.028	19.39	379	5.33
4,000	0.480	0.027	4,400	0.025	20.79	311	3.78
4,800	0.497	0.023	5,600	0.019	22.15	242	2.31
6,400	0.510	0.017	7,200	0.015	22.82	199	1.48
8,000	0.515	0.014	8,000	0.014		0	0.00

$$r = \frac{108}{p} \qquad\qquad S_w = 0.0187 \sum \bar{p}\Delta V$$

4.18 Illustrative Examples

4.18.1 Narrow size range

The curve of intruded volume , against pressure p is called the pressurizing curve, and from it the various distributions can be found. If the pore size distribution is narrow it is possible to plot the cumulative pore size distribution by volume on linear graph paper and to differentiate the curve to obtain the relative pore size distribution by volume (Table 4.2). The data are plotted as a cumulative and relative pore size distribution by volume in Figures 4.10.and 4.11 and by surface in Figures 4.12 and 4.13.

4.18.2 Wide size range

For a wide distribution it is preferable to plot the distributions on log–linear paper with V on the linear axis. The pressurizing curve is shown in Figure 4.14. Data extracted from the pressurizing curve are presented in Table 4.3. The data are plotted as cumulative and relative size distribution by volume in Figures 4.15 and 4.16 and the

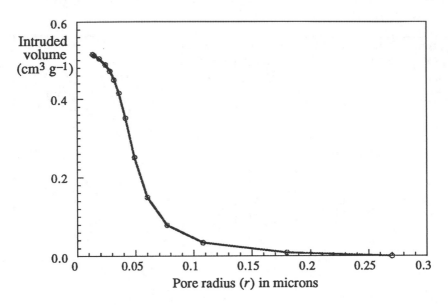

Fig. 4.10 Cumulative pore size distribution by volume by mercury porosimetry for a sample having a narrow pore size range.

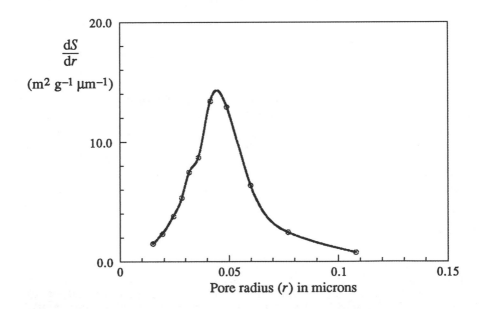

Fig. 4.11 Relative pore size distribution by volume by mercury porosimetry for a material having a narrow pore size distribution.

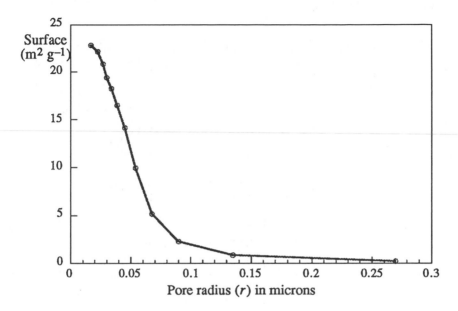

Fig. 4.12 Cumulative pore size distribution by surface by mercury porosimetry for a sample having a narrow pore size distribution.

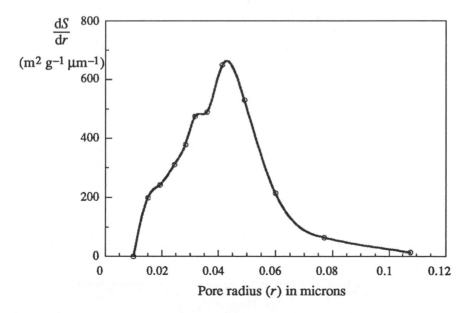

Fig. 4.13 Relative pore size distribution by surface by mercury porosimetry for a material having a narrow pore size range.

Table 4.3 Evaluation of pore size distribution by mercury porosimetr
(wide size range)

Pressure (p) psia	Volume intruded (V) cm^3 g^{-1}	Pore radius (r) μm	Mean pore radius ($\bar{r}$) μm	Specific surface S_W m^{2}g^{-1}	Relative volume $\dfrac{\Delta V}{\Delta(\log r)}$ cm^3 g^{-1} log μm^{-1}	Relative surface $\dfrac{\Delta S}{\Delta(\log r)}$ m^{2}g^{-1}log μm^{-1}
1.00	0.000	108	58.1	0.00	0.01	0.235
2.72	0.102	39.7	33.5	0.00	0.01	0.088
3.72	0.114	29.0	19.4	0.01	0.01	0.080
7.41	0.138	14.6	7.85	0.01	0.01	0.032
20.1	0.152	5.37	2.89	0.02	0.03	0.037
54.7	0.168	1.97	1.06	0.07	0.11	0.055
148.7	0.192	0.726	0.391	0.28	0.50	0.097
404.0	0.234	0.267	0.202	0.70	1.93	0.193
667.0	0.276	0.162	0.122	1.58	4.03	0.244
1,100.0	0.329	0.0982	0.074	2.75	5.40	0.198
1,813.0	0.372	0.0596	0.045	4.37	7.45	0.166
2,987.0	0.408	0.0362	0.027	6.51	9.88	0.134
4,925.0	0.437	0.0219	0.017	9.56	14.04	0.115
8,120.0	0.462	0.0133	0.010	15.19	25.92	0.129
13,390.0	0.490	0.0081	0.0061	27.46	56.53	0.170
22,070.0	0.527	0.0049	0.0037	64.63	171.1	0.313
36,390.0	0.595	0.0030	0.0022	133.85	318.7	0.354
60,000.0	0.672	0.0018				

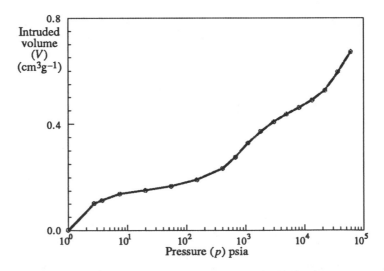

Fig. 4.14 Pressurizing curve for a wide pore size distribution sample.

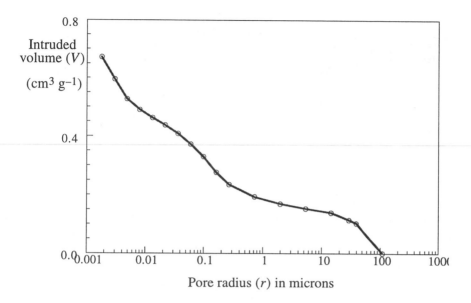

Fig. 4.15 Cumulative pore size distribution by volume for a sample with a wide range of pore sizes.

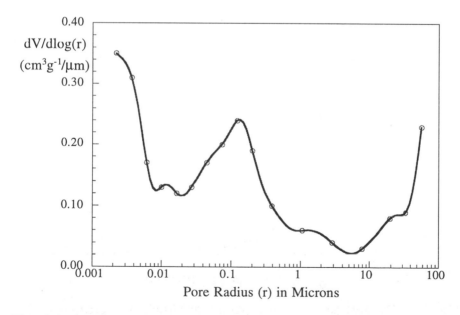

Fig. 4.16 Relative pore volume frequency for sample with a wide range of pore sizes

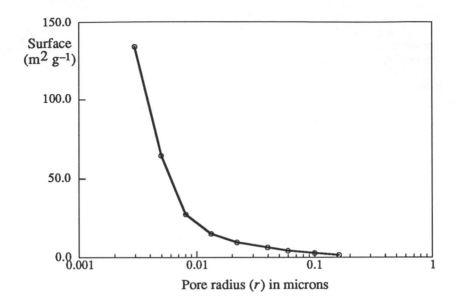

Fig. 4.17 Cumulative pore size distribution by surface by mercury porosimetry for sample having a wide pore size range.

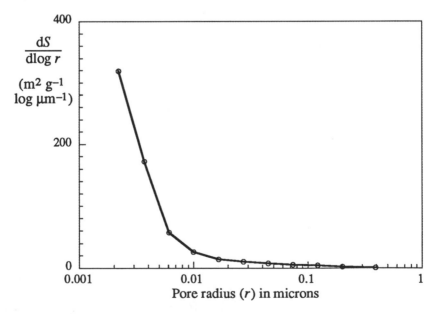

Fig. 4.18 Relative pore size distribution by surface by mercury porosimetry for sample having a wide pore size range.

cumulative and relative size distribution by surface in Figures 4.17 and
4.18. In this case the frequency curve is on the basis of frequency per
log (μm) so that the areas under the curves are the total volume and
surface respectively. Tabular data from a pressurizing curve may
contain 50 or more experimental points and a tabular differentiation
can produce a noisy curve; the noise may justifiably be attributed to
experimental error; in such cases a graphical differentiation is
preferred.

4.19 Commercial equipment

Mercury porosimeters are manufactured by Carlo Erba Fisons
Micromeretics and Quantachrome Corporation and operate in
incremental, continuous and dual mode.

4.19.1 Incremental mode

The principle of the incremental technique may be understood by
referring to Figure 4.19. The sample is degassed to remove adsorbed
vapors. The time of degassing depends upon the material to be tested
and can be greatly reduced if the material is oven dried before testing.
Mercury is then introduced until it completely covers the sample and
any excess is drained off. In the incremental mode the pressure is
increased in increments and the system allowed to come to equilibrium
before the intruded volume of mercury is determined. The pressure is
raised to 0.5 psia from which point the analysis begins. The pressure is
raised manually or automatically in steps of about 1psia to atmospheric
pressure. After each increment a reading is made of the volume of
mercury penetrating into the sample. Next the chamber around the
sample cell is filled with hydraulic fluid and the pressure increased.

The Carlo Erba 2000 operates in the 0–20 MPa pressure range to
measure pore diameters down to 7.5 nm. As the pressure is increased
in steps from vacuum to one atmosphere, macropores with diameters
in excess of 15 μm are measured. The dilatometer is then placed in a
high pressure unit in order that mesopores can be measured. The
mercury level in the dilatometer stem is electrically followed by means
of a capacitive system. The instrument has a programmable delay
which permits the system to come to equilibrium at each step.

4.19.2 Incremental and continuous mode

The Micromeretics AutoPore II 9220 measures up to six samples at a
time in four low pressure and two high–pressure stations. This permits
either continuous or step wise operation; the latter with up to 250 steps,
covering the pore diameter ranges 360 to 6 μm (low pressure), 6 to
0.003 μm (high pressure).

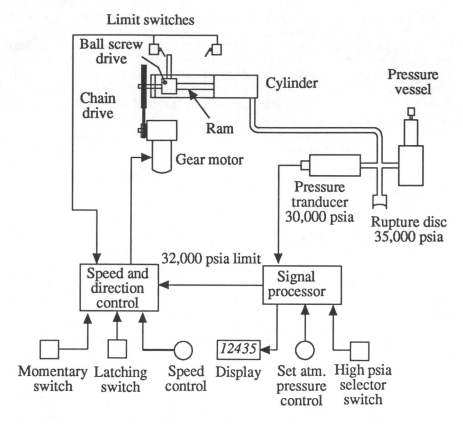

Fig. 4.19 Micromeretics 9320 high pressure generation and control system.

The Micromeretics Poresizer 9320 has improved graphics and resolution with user friendly, menu driven software. The sample cell, or penetrometer (Figure 4.4) consists of two pieces, (a) the portion that contains the sample and (b) the cap with the precision bore tubing that contains the dilatometer. The two main parts are fitted together by means of ground surfaces. A weighed sample of powder is placed in the sample space; part (b) is then fitted on to part (a), and the sample holder is placed in the sample chamber. The pressure chamber is a heavy walled steel vessel since it has to withstand very high pressure.

4.19.3 Continuous mode

The continuous (scanning) technique was pioneered by Quantachrome, who offer three Autoscan porosimeters.

The Autoscan 60 operates over the pressure ranges (psia): 0.5–24, ambient to 6000 and ambient to 60,000; an optional range of ambient to 1200 psia is also available. Intrusion and extrusion volumes are plotted as a continuous curve on an X–Y recorder. Ultra–high resolution of pore distributions is provided by obtaining up to 1700 data points along the curve. A separate filling apparatus (Figure 4.5) is used to evacuate the sample cell, outgas the sample, fill the sample cell with mercury and perform intrusion/extrusion runs from 0.5 psia to 24 psia (approximately 200 to 4 μm diameter). If sub-ambient data is not required an alternative low cost filling apparatus is available.

The Autoscan 33 operates over the pressure ranges (psia) 0.5–24, 20– 3300 and 20 to 33,000 psia (0.003 to 213 μm diameter).

These two instruments are designed to compensate for mercury compressibility thermodynamically by balancing the slight expansion due to compressive heating and the associated change in the hydraulic oil dielectric value with pressure. A cell filled with mercury will indicate a deviation on the volume axis of less than 0.5% of full scale therefore there is no need for a blank run.

The Autoscan 500 operates over the pressure ranges (psia) 0.5–50 and 0.5–500 to cover pore diameters from 213 to 0. 2 μm. It is used for macroporous material and can accommodate samples up to 20 cm^3 in volume with a maximum pore volume of 2 cm^3.

The rate at which the pressure is generated is continuously variable in all Autoscan porosimeters and the maximum pressure can be generated in as little as 5 min or it can take over 5 h for samples requiring lower intrusion rates. Thus, under favorable conditions, a trained operator can produce up to six complete porosimetry analyses per hour.

Fisons Pascal Scrics gcncratcs a high resolution penetration curve in as little as 9 min. The correct pressurization speed is determined automatically according to the presence or absence of pores. It is claimed that other pressurization techniques such as continuous scanning, predefined pressure matrix or stepwise pressurization need more than 30% more time. Pascal 140 is a fully automatic low pressure (0–400 kPa) porosimeter. Pascal 2240 is a more versatile instrument operating at pressures up to 200 MPa. Its large volume autoclave and special capacitive detection system accepts dilatometers of different sizes, thus permitting analysis of a wide range of samples. Pascal 440 provides high resolution and analysis speeds at pressures up to 400 MPa.

4.20 Anglometers

Wetting angle measurements have been reported using both static and dynamic techniques. Static measurements confirm that the advancing angle differs from the receding angle. In order to conform with the Washburn equation it is necessary to determine the advancing angle for mercury. In the Quantachrome Anglometer the pressure required to

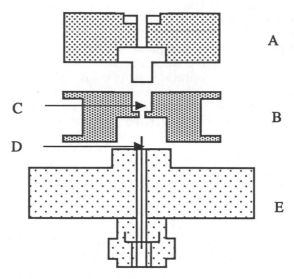

Fig. 4.20 Connection apparatus (assembly diagram). A, die; B, sample holder; C, sample cup; D, 0.8 mm steel pin; E, pin holder.

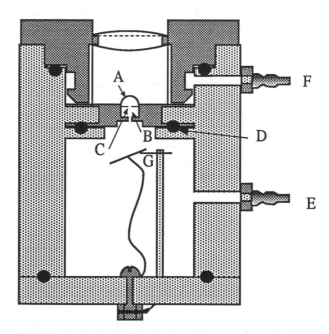

Fig. 4.21 Cell housing. A, mercury head; B, powder bed; C, hole in powder bed; D, O–ring seal; E, vacuum port; F, vacuum and vent port; G, piezoelectric detector.

break through an 800 μm hole drilled through a powder sample is determined. A finely powdered sample is placed in the cup C (Figure 4.20) contained within a sample steel holder B through which an 800 μm pin is extended. A die A is placed over the pin on top of the powder and the assembled compaction apparatus is placed in a hand jack. The powder is compacted at such a pressure (< 100 psia) that a clean hole is made when the pin is removed. The sample holder is then placed in the cell housing (Figure 4.21) of the anglometer and a known volume of triple distilled mercury A is added on top of the powder bed, resulting in a known pressure head. Pressure is then applied and when the mercury breaks through the hole it impinges on a piezoelectric detector G, which provides a signal for a pressure display with a resolution of 0.001 psia. The Washburn equation is then applied in order to determine γ.

4.21 Assessment of mercury porosimetry.

4.21.1 Effect of experimental errors

Figure 4.22 shows the relative pore distribution using narrow size intervals and using a curve smoothing operation. When one considers the uncertainties in the data it becomes probable that the sharp peaks are not real but due to experimental errors. Figure 4.23 gives a typical presentation of data from a mercury porosimeter.

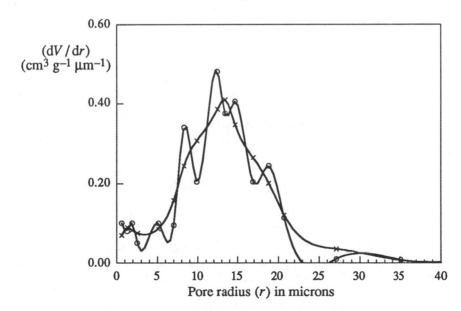

Fig. 4.22 Effect of smoothing out experimental errors.

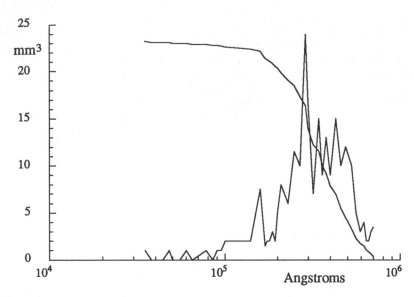

Fig. 4.23 Direct plotting of data from a mercury porosimetry analysis.

In a comparison between surface area determination of ceramics by nitrogen adsorption and mercury porosimetry it was found that the results were similar in the surface area range $10–100$ m^2 g^{-1} but outside this range deviations were significant [78].

4.22 Mercury porosimetry report

Table 4.4 shows a typical mercury porosimetry data sheet for a magnesium hydroxide having a wide size range. Column 1 shows the pressure and column 2 the intruded volume corrected for compressibility by means of a blank run. From these data the pressurizing curve, Figure 4.24, is generated. Conversion of pressure into pore radius (column 3) permits determination of the cumulative pore size distribution by volume (Fig. 4.25). The cumulative pore size distribution by surface is calculated in column 4 and the data presented in Figure 4.27. The mean pore radii are presented in column 5 and the log differentials of the volume and surface data are presented in columns 6 and 7 and Figures 4.28 and 4.29.

The total intruded volume is 1.3033 cm^3 g^{-1} and the total surface is 66.15 m^2 g^{-1}. Figure 4.26 indicates some inter–particle filling above 0.5 μm, pores in the size range 0.1 to 0.5 μm and micropores in the size range 0.007 to 0.1 μm.

An examination of the data shows that 6% of the mercury intruded into pores smaller than 0.007 μm but that these micropores contributed 59% of the surface. 18% of the mercury intruded into pores in the size

Table 4.4 Mercury intrusion data for magnesium hydroxide

Pressure (p) psia	Intruded volume (V) cm^3 g^{-1}	Pore radius (r) μm	Cumulative surface (S) m^{2}g^{-1}	Mean radius ($\bar{r}$) μm	$\dfrac{dV}{d\log r}$ (cm^3 g^{-1} μm^{-1})	$\dfrac{dS}{d\log r}$ (m^{2}g^{-1} μm^{-1})
5.97	0.0210	15.151	0.00	16.639	0.2697	0.03
6.97	0.0404	12.977	0.01	14.064	0.2884	0.04
7.96	0.0568	11.363	0.01	12.170	0.2843	0.05
8.96	0.0706	10.095	0.01	10.729	0.2685	0.05
9.94	0.0821	9.0996	0.01	9.5972	0.2551	0.05
11.94	0.1032	7.5754	0.02	8.3375	0.2650	0.06
12.93	0.1124	6.9954	0.02	7.2854	0.2659	0.07
14.93	0.1271	6.0583	0.03	6.5268	0.2353	0.07
16.92	0.1400	5.3457	0.03	5.7020	0.2374	0.08
19.91	0.1557	4.5429	0.04	4.9443	0.2222	0.09
21.91	0.1636	4.1283	0.04	4.3356	0.1900	0.09
24.90	0.1764	3.6325	0.05	3.8804	0.2304	0.12
26.96	0.1781	3.3550	0.05	3.4938	0.0492	0.03
30.39	0.1873	2.9763	0.05	3.1656	0.1769	0.11
40.39	0.2057	2.2394	0.07	2.6079	0.1489	0.11
60.19	0.2326	1.5027	0.10	1.8711	0.1553	0.17
79.60	0.2582	1.1363	0.13	1.3195	0.2109	0.32
100.34	0.2786	0.9014	0.17	1.0189	0.2029	0.40
124.48	0.2979	0.7266	0.22	0.8140	0.2061	0.51
149.66	0.3153	0.6044	0.27	0.6655	0.2175	0.65
174.86	0.3324	0.5173	0.34	0.5608	0.2530	0.90
199.82	0.3508	0.4527	0.41	0.4850	0.3175	1.31
249.48	0.6034	0.3626	1.65	0.4076	2.6204	12.86
301.90	0.7877	0.2996	2.76	0.3311	2.2251	13.44
351.75	0.8590	0.2571	3.28	0.2784	1.0743	7.72
401.74	0.8971	0.2251	3.59	0.2411	0.6602	5.48
450.32	0.9233	0.2009	3.84	0.2130	0.5285	4.96
499.49	0.9401	0.1811	4.01	0.1910	0.3733	3.91
554.36	0.9588	0.1632	4.23	0.1721	0.4131	4.80
599.16	0.9676	0.1510	4.34	0.1571	0.2607	3.32
650.71	0.9762	0.1390	4.46	0.1450	0.2399	3.31
699.54	0.9827	0.1293	4.56	0.1342	0.2068	3.08
748.36	0.9889	0.1209	4.66	0.1251	0.2116	3.38
799.60	0.9948	0.1131	4.76	0.1170	0.2051	3.51
852.77	0.9997	0.1061	4.85	0.1096	0.1753	3.20
901.27	1.0050	0.1004	4.95	0.1032	0.2206	4.28
950.10	1.0082	0.0952	5.02	0.0978	0.1397	2.86
1,002.8	1.0118	0.0902	5.09	0.0927	0.1536	3.31
1,255.6	1.0288	0.0720	5.51	0.0811	0.1739	4.29

Table 4.4 (continued) mercury intrusion data for magnesium hydroxide

Pressure (p) psia	Intruded volume (V) cm^3 g^{-1}	Pore radius (r) μm	Cumulative surface (S) m^{2}g^{-1}	Mean radius ($\bar{r}$) μm	$\dfrac{dV}{d\log r}$ (cm^3 g^{-1} μm^{-1})	$\dfrac{dS}{d\log r}$ (m^{2}g^{-1} μm^{-1})
1,504.5	1.0412	0.0601	5.89	0.0661	0.1581	4.79
1,749.4	1.0545	0.0517	6.36	0.0559	0.2030	7.26
2,002.8	1.0640	0.0452	6.76	0.0484	0.1617	6.68
2,993.5	1.0977	0.0302	8.55	0.0377	0.1931	10.25
4,976.2	1.1633	0.0182	13.97	0.0242	0.2972	24.57
6,971.3	1.2177	0.0130	20.95	0.0156	0.3715	47.71
9,969.1	1.2497	0.0091	26.76	0.0110	0.2060	37.37
12,445	1.2583	0.0073	28.86	0.0082	0.0893	21.85
14,939	1.2624	0.0061	30.09	0.0067	0.0517	15.51
32,338	1.2768	0.0028	37.64	0.0029	0.0875	60.11
34,848	1.2782	0.0026	38.68	0.0027	0.0433	32.08
37,401	1.2796	0.0024	39.80	0.0025	0.0455	36.26
39,874	1.2821	0.0023	41.93	0.0023	0.0899	76.72
42,358	1.2830	0.0021	42.75	0.0022	0.0343	31.16
44,835	1.2859	0.0020	45.54	0.0021	0.1175	113.16
47,276	1.2875	0.0019	47.17	0.0020	0.0695	70.73
49,768	1.2898	0.0018	49.63	0.0019	0.1031	110.50
54,735	1.2948	0.0017	55.40	0.0017	0.1210	139.53
59,930	1.3033	0.0015	66.15	0.0016	0.2158	273.04

Data: $\gamma = 0.485$ N m^{-1} $\theta = 130°$ hence $r = \dfrac{0.624}{p}$

With p in psia and r in μm, $r = \dfrac{90.43}{p}$

range 0.007 to 1 μm contributing 33% of the surface. Inter-particle filling used up 76% of the intruded mercury and the superficial surface of 5 cm^3 g^{-1} made up 8% of the total surface.

4.23 Liquid porosimetry

The high pressures used in mercury porosimetry gives erroneous results with deformable material such as fabric. A more general version of liquid porosimetry, based on the same principles, uses a variety of liquids and can be carried out in the extrusion mode [79,80]. In this technique, a pre–saturated specimen is placed on a microporous membrane supported on a rigid porous plate in an enclosed chamber. The gas pressure within the chamber is increased in steps causing the liquid to flow out of the pores. The amount of liquid removed is monitored by a top–loading recording balance. One also has the

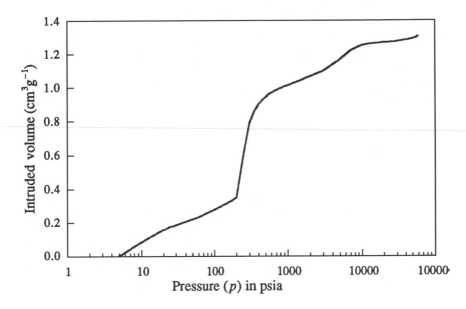

Fig. 4.24 Pressurizing curve for a powder having a wide pore size range.

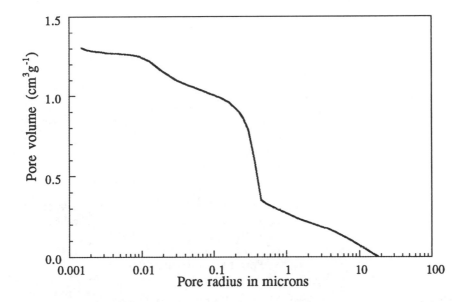

Fig. 4.25 Cumulative pore size distribution by volume.

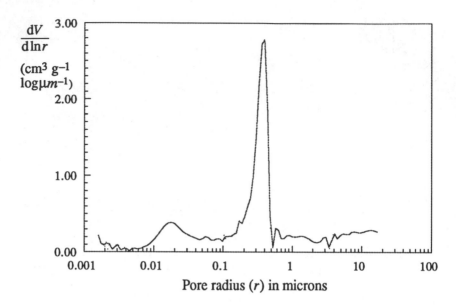

Fig. 4.26 Relative pore size distribution by volume (smoothed curve).

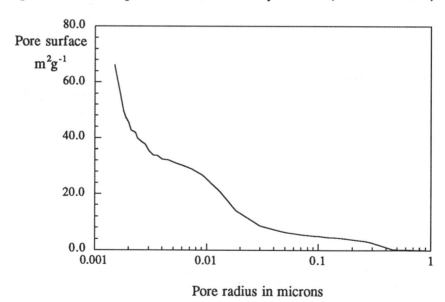

Fig. 4.27 Cumulative pore size distribution by surface.

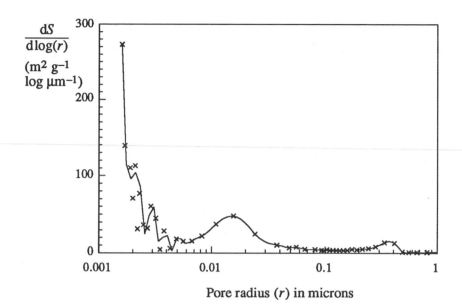

Fig. 4.28 Relative pore size distribution by surface.

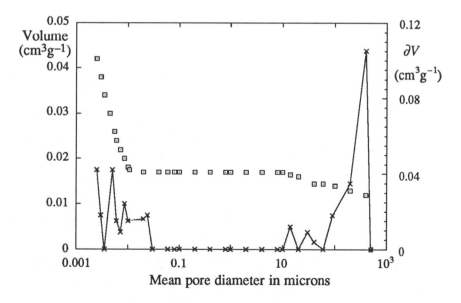

Fig. 4.29 Raw intrusion data for a polymer.

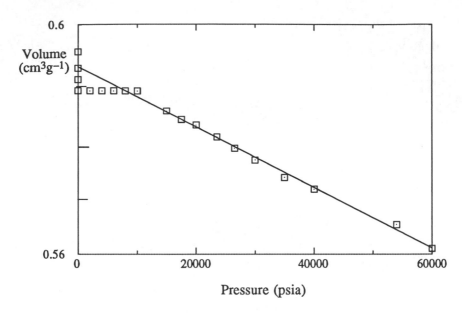

Fig. 4.30 Compressibility curve for a non–porous polymer.

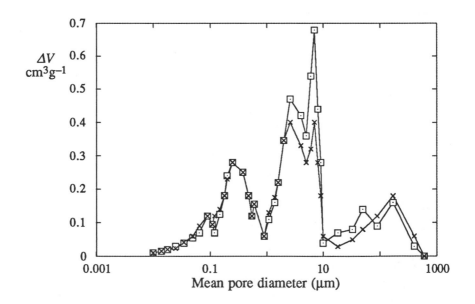

Fig. 4.31 Independent measurements of pore size distribution by two laboratories.

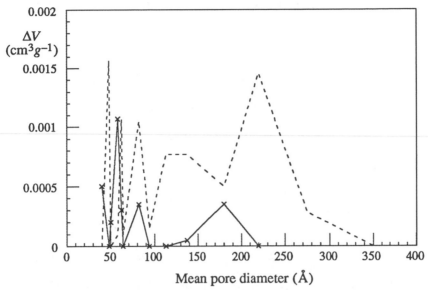

Fig. 4.32 Effect of stressing on the microstructure of a polymer.

option of starting with a dry material at elevated pressure and allowing liquid uptake as the pressure is reduced. It is also possible to determine contact angles within porous networks by comparing uptake versus applied pressure using the liquid of interest and a reference liquid with a known contact angle [81] . A fully computer controlled liquid porosimeter has been developed and a description of its mode of operation given [82].

4.24 Applications

Figure 4.29 shows the mercury intrusion curve for a polymer. Inter-particle filling takes place (void diameters between 10 and 1000 μm) initially, followed by a plateau region and finally apparent micropores in the sub-0.02 μm region. A blank run is shown in Figure 4.30 and it can be seen that the system undergoes compressibility at pressures above 10,000 psia. After correcting for this the apparent micropores disappear. Data for this, and the following illustrations were provided by Dr, Lloyd Abrams, manager of the Corporate Catalyst Characterization Center at DuPont Experimental Station, Wilmington. Figure 4.30 shows the pore size distribution of a metal substrate. The initial intrusion at low pressure is due to penetration into interparticle voids; as the pressure is increased, the voids within the substrate are penetrated, finally the micropores within the metal are filled. The reproducibility of micropore data (0.01 to 1 μm diameter) generated at two different sites, is remarkable. The lack of agreement for the larger pores indicates that the two samples were packed differently. Finally,

Figure 4.32 shows how stressing a polymer generates micropores in the material.

References

1   Zweitering, P. (1958), *The Structure and Properties of Porous Solids,* Butterworths, *149*
2   Joyner, L.G., Barrett, E.P. and Skold, R. (1951), *J. Am. Chem. Soc.*, **73**, 3158, *149*
3   Adamson, A.W. (1967), *Physical Chemistry of Surfaces,* Interscience, N.Y., *150*
4   Washburn, E.W. (1921), *Proc. Natn. Acad. Sci.*, **7**, 115–6, *151, 154*
5   Henderson, L.M., Ridgeway, C.M. and Ross, W.B. (1940), *Refin. Nat. Gas Mfr.*, **19**(6), 69–74, *155*
6   Ritter, H.L. and Drake, L.C. (1945), *Ind. Engng. Chem., Anal. Edn.*, **17**, 782–786, *155, 157*
7   Ritter, H.L. and Drake, L.C. (1945), *Ind. Engng. Chem., Anal. Edn.*, **17**, 787–791, *155, 157*
8   Drake, L.C. (1949), *Ind. Engng. Chem*, **41**, 780–785, *155*
9   Burdine, N.T., Gourney, L.S. and Reichertz, P.P. (1950), *Trans. Am. Inst. Min. Metal. Engrs.*, **189**, 195–204, *155*
10 Bucker, H.P., Felsenthal, M. and Conley, F.R. (1956), *J. Petrol. Technol.*, AIME, 65–66, *155*
11 Winslow, N.M. and Shapiro, J.J. (1959), *ASTM Bull.*, TP49, 39–44, 107, *155, 157*
12 Guyer, A., Boehlen, B. and Guyer, A.S. Jr (1959), *Helv. Chim. Acta,* **42**, 2103–2110, *155*
13 McKnight, T.S., Marchassault, R.H. and Mason, F.G. (1958), *Pulp Paper Mag. Can.*, **59**(2), 81–88, *155*
14 Purcell, W.R. (1949), *J. Petrol. Technol.*, **1**, 39–48, *155*
15 Stromberg, R.R. (1955), *J. Res. NBS*, **54**, 73–81, *155*
16 Watson, A., May, J.O. and Butterworth, B. (1957), *Trans. Br. Ceram. Soc.*, **56**, 37–50, *155*
17 Plachenov, T.G. (1955), *J. Appl. Chem. USSR,* **28**, 223, *155*
18 Dullien, F.A.L. and Dhawwan, G.K. (1976), *J. Colloid Interf. Sci.*, **52**(1), 573–575, *155*
19 Svata, M. (1968), *Abh. Sächs Akad. Wiss.*, **49**(5), 191–196, *155*
20 Svata, M. and Zabransky, Z. (1968/69), *Powder Technol.*, **2**, 159–161, *155*
21 Baker, D.J. (1971), *J. Phys. E*, 4(5), 388–389, *155*
22 Leppard, C.J. and Spencer, D.M.T. (1968), *J. Phys. E*, **1**, 573–575, *155*
23 Reich, B. (1967), *Chem. Ing. Tech.*, **39**(22), 1275–1279, *156*
24 Meyer, H.I. (1953), *J. Appl. Phys.*, **24**, 510–512, *156*
25 Zgrablich, G., Mendioroz, S. and Daza, L. (1991), *Langmuir,* 7(4), 779–785, *156*

26   Zhdanov, V.P. and Fenelonov, V.B. (1987), *React. Kinet. Catal. Lett.*, **33**(2), 377–380, *156*
27   Tsetsekov, A., Androotsopoulos, G. and Mann, R. (1991), *Chem. Eng. Commun.*, **110**, 1–29, *156*
28   Frevel, L.K. and Kressley, L.J. (1963), *Anal. Chem.*, **35**, 1492, *156*
29   Meyer, R.P. and Stowe, R.A. (1965), *J. Colloid Sci.*, **20**, 893, *156*
30   Svata, M. and Zabransky, Z. (1969/70), *Powder Technol.*, **3**, 296–298, *156*
31   Rootare, H.M. and Craig, R.G. (1974), *Powder Technol.*, **9**, 199–211, *156*
32   Conner, W.C. Blanco, C, Coyne, K, Neil, J., Mendioroz, S. and Pajares, J. (1987), *Stud. surf. sci. catal.*, **39**, 273–281, *156*
33   Reverberi, A., Ferraiolo, G. and Peloso, A. (1966), *Ann. Chim.*, **56**, 1552, *156*
34   Svata, M. (1971/72), *Powder Technol.*, **5**, 345–349, *156*
35   Adkins, B.D., Heink, J.B. and Davis, B.H. (1987), *Adsorpt. Sci. Technol.*, **4**(1–2), 87–104, *157*
36   Johnson, M.F.L. (1988), *J. Catal.*, **110**(2), 419–422, *157*
37   Whittemore, O.J. Jr and Sipe, J.J. (1974), *Powder Technol.*, **9(4)**, 159–164, *157*
38   Gillard, J. (1975), *Lab–Pharma–Probl. Tech.*, **23**, 246, 789–799, *157*
39   Rappeneau, J. (1965), *Les Carbenes*, 2, Masson, Paris, Ch. 14, pp. 134–140, *157*
40   Scholten, J.J.F. (1967), *Porous Carbon Solids*, ed. R.L. Bond, Academic Press, pp. 225–249, *157*
41   Diamond, S. (1970), *Clay Min.*, **18**, 7, *157*
42   Rootare, H.M. (1970), *Adv. Exp. Techn. in Powder Metall.*, **V**, Perspectives in Powder Metall., Plenum Press, pp. 225–254, *148, 157, 162*
43   Hearn, H. and Hooton, R.D. (1992), *Cem. Concr. Res.*, **2**(5), 970–980, *157*
44   Rootare, H.M. (1968), *Aminco Isb. News*, **24**(3), *155, 157*
45   Wardlow, C.N. and McKellar, M., (1981), *Powder Technol.*, **29**, 127, *159*
46   Wardlow, C.N. and Li, Y. (1988), *Transparent Porous Media*, **3**, 17, *159*
47   Ioannidis, M.A., Chatzis, I. and Payatakes, A.C., (1991), *J. Colloid Interf. Sci.*, **143**, 22, *159*
48   Chatzis, I., Morrow, N.R. and Lim, H.T., (1983), *Soc. Pet. Eng.*, 311, *159*
49   Ioannidis, M.A. and Chatzis, I., (1993), *Hysteresis and Entrapment from Porosimetry*, Academic Press, *159*
50   Meyer, H.I. (1953), *J. Appl. Phys.*, **24**, 510–512, *159*
51   Cohrt, H. and Proz, H. (1981), *Powder Metall. Int.* 3, 121–5, *160*
52   Liabastre, A.A. and Orr, C. (1978), *J. Colloid Interf. Sci.*, **64**, 1, 1–18, *161*

53   Kloubek, J. (1993), *Carbon,* **31,** 445, *161*
54   Kloubek, J. (1994), *J. Colloid Interf. Sci.,* **163,** 10–18, *161*
55   Winslow, D.N. and Diamond, S. (1970), *J. Mater.,* **5,** 564, *161*
56   Juhola, A.J. and Wiig, O (1949), *J. Am. Chem. Soc.,* **71,** 2078–2080, *161*
57   Winslow, D.N. (1978), *J. Colloid Interf. Sci.,* **67,** 42, *161*
58   Kloubek, J. (1992), *J. Adhesion Sci. Technol.,* **6,** 667, *161*
59   Guo, X., Wang, B.and Ho, B. (1990), *Lizi, Jiahuan Yu Xifu,* 6(2), 87–92 (Chinese), *161, 162*
60   Wit, L.A. de and Scholten, J.J.F. (1975), *J. Catalysis,* **36,**36, *161*
61   Good, R.J. and Mikhail, R.Sh., *Powder Technol.,* *161*
62   Lowell, S. and Shields, J.E. (1981), *J. Colloid. Interf. Sci.,* **80,** 192 39, *162*
63   Lowell, S. and Shields, J.E. (1981), *J. Colloid. Interf. Sci.,* **83,** 273 39, *162*
64   Heertjes, P.M. and Kossen, N.W.F. (1967), *Powder Technol.,* **1,** 33–42, *162*
65   Brockel, U. and Löffler, F. (1991), *Part. Part. Charact.,* **8,** 215–221, *162*
66   Rootare, H.M. and Prenzlow, C.F. (1967), *J. Phys. Chem.,* **71,** 2734–2736, *158, 162*
67   Roberts, N.K. (1964), *J. Chem. Soc.,* 1907–1915k, *162*
68   Kloubek, J. (1981), *Powder Technol.,* **29,** 63, *162*
69   Spencer, D.T.H. (1969), *Brit. Coal Util. Res. Assoc. Monthly Bull.,* **33,** 22, 8, *163, 165*
70   Scholten, J.J.F. (1967), *Porous Carbon Solids,* Academic Press, London, *163*
71   Lee, J.A. and Maskell, W.C. (1973), *Powder Technol.,* **7,** 259–262, *113, 165*
72   Palmer, H.K. and Rowe, R.C. (1974), *Powder Technol.,* **9,** 181–186, *165*
73   Baker, D.J. and Morris, J.B. (1971), *Carbon,* **9,** 687, *165*
74   Feldnan, R.F. (1984), *J. Am. Chem. Soc.,* **67,** 30, *165*
75   Koubek, J. (1994), *J. Colloid Interf. Sci.,* **163,** 10–18, *165*
76   Winslow, D.N. and Diamond, S., (1970), *J. Mater.,* **5,** 564, *165*
77   Lowell, S. and Shields, J.E., (1983), *Advances in the Mechanics and theFlow of Granular Materials,* 1, p. 194, Trans. Tech. Publ., *165*
78   Milburn, D.R. and Davis, B.H. (1993), *Ceram. Eng. Sci. Proc.,* **14**(11–12), 130–134, *180*
79   Miller, B. and Tyomkin, I. (1983), *Proc. 11th INDA Technical Symp.on Nonwoven Technology,* p. 73, *182*
80   Miller, B. and Tyomkin, I. (1986), *Textile Res. J.,* **56,** 35, *182*
81   Klein, D.M., Tyomkin, I., Miller, B. and Rebenfeld, L. (1990), *J. Appl. Bio. Materials,* **1,** 137, *187*
82   Miller, B. and Tyomkin, I. (1994), *J. Colloid Interf. Sci.,* **162,** 163–170, *187*

5

Other methods for determining surface area

5.1 Introduction

Nitrogen gas adsorption at liquid nitrogen temperature is the most widely used method of surface area determination [1]. Krypton at liquid nitrogen temperature is favored for low surface area powders. Most gases can and have been used and these include water vapor at room temperature and at 78°C. The problems that arise when one deviates from standard conditions are the choice of the applicable molecular area for the adsorbate and the correct theoretical model to use. The first question is usually resolved by using published values or carrying out experiments to determine molecular area by comparison with standard BET procedures using nitrogen at liquid nitrogen temperature. Since there is no unanimity in published literature the second method is probably preferable. When coverage is very low, as with carbon dioxide at room temperature, the Freundlich equation may be applicable.

Permeametry is often used for control purposes due to its simplicity [2]. Surface areas may also be calculated from size distribution data [3].

Other adsorption techniques include adsorption from solution; here, one of the problems is that of the accurate determination of the very small amounts adsorbed. Adsorption studies have been carried out with fatty acids, polymers, ions, dyestuffs and electrolytes using a range of analytical techniques. The usual way of determining a single point on the adsorption isotherm of a binary solution is to bring a known amount of solution of known concentration into contact with a known weight of adsorbent in a vessel at the required temperature, and agitate for several hours. After equilibrium has been reached an aliquot part of the bulk solution is withdrawn and the concentration change determined by some suitable method. The amount adsorbed can then be determined and related to powder surface area.

Surface areas may also be determined from heats of adsorption and the technique has been greatly simplified with the introduction of sensitive flow micro-calorimeters. These can be used with liquid or gas mixtures to determine both heats of adsorption and amount adsorbed: the method thus provides information on molecular areas and orientations as well as energies of adsorption.

5.2 Determination of specific surface from size distribution data

5.2.1 Number distribution

Let particles of size $d_{x,r}$ constitute a fraction m_r of the total number N in a powder so that:

$$m_r = \frac{n_r}{N} \text{ where } N = \Sigma n_r$$

The surface area of the powder is given by:

$$S = \alpha_{s,x} N \sum m_r d_{x,r}^2$$

The volume is:

$$V = \alpha_{v,x} N \sum m_r d_{x,r}^3$$

The mass specific surface is:

$$S_w = \frac{S}{\rho V} = \frac{\alpha_{sv,x}}{\rho} \frac{\displaystyle\sum_{r=0}^{\max} m_r d_r^2}{\displaystyle\sum_{r=0}^{\max} m_r d_r^3} \tag{5.1}$$

x refers to the method of measurement being A for a sieve analysis St for a Stokes analysis and so on, [e.g. d_A is the sieve diameter and d_{St} the Stokes diameter. α refers to the shape coefficients, the suffixes denoting surface (s), volume (v) and surface-volume (sv)].

For a system of spheres:

$$S = \pi N \sum m_r d_r^2; \quad V = \frac{\pi}{6} N \sum m_r d_r^3; \quad S_V = 6 \frac{\sum m_r d_r^2}{\sum m_r d_r^3} \tag{5.2}$$

The mass specific surface S_w is related to the volume specific surface by $S_w = \rho S_v$ where ρ is the density of the powder. Assuming a value of $\alpha_{sv} = 6$ gives, for example, the mass specific surface of a powder by microscopy:

$$S_{w,a} = \frac{6}{\rho} \frac{\sum\limits_{r=0}^{\text{max}} m_r d_r^2}{\sum\limits_{r=0}^{\text{max}} m_r d_r^3} \qquad (5.3)$$

5.2.2 Surface distribution

Let particles of size $d_{x,r}$ constitute a fraction t_r of the total surface S in a powder so that:

$$t_r = \frac{S_r}{S} \text{ where } S = \Sigma S_r$$

$$St_r = \alpha_{s,x} n_r d_{x,r}^2$$

$$V_r = \alpha_{v,x} n_r d_{x,r}^3$$

$$St_r = \frac{\alpha_{s,x}}{\alpha_{v,x}} \frac{V_r}{d_{x,r}}$$

$$S = \frac{\alpha_{sv}}{\sum\limits_{r=0}^{\text{max}} t_r d_{x,r}} \qquad (5.4)$$

Assuming α_{sv} is constant over the size range being measured.

5.2.3 Volume (mass) distribution

Let particles of size $d_{x,r}$ constitute a fraction q_r of the total volume V in a powder so that:

$$q_r = \frac{V_r}{V} \text{ where } V = \Sigma V_r$$

$$Vq_r = \alpha_{v,x} n_r d_{x,r}^3$$

$$S_r = \alpha_{s,x} n_r d_{x,r}^2$$

$$S_v = \alpha_{sv} \sum\limits_{x=0}^{\text{max}} \frac{q_r}{d_{x,r}} \qquad (5.5)$$

Example (Specific surface area determination from sieve analysis)

Sieve aperture size d_{Ar} (μm)	Mass percentage undersize Σq_r	Mean sieve size $\bar{d}_{A,r}$ (μm)	Mass fraction between sieve sizes q_r	$\Sigma \dfrac{q_r}{\bar{d}_{A,r}}$
53.0	0	–	–	–
63.0	0.20	57.80	0.001953	0.000034
75.0	1.95	68.73	0.017578	0.000290
89.1	8.98	81.74	0.070313	0.001150
106.0	25.39	97.20	0.164063	0.002838
126.1	50.00	115.59	0.246094	0.004967
149.9	74.61	137.46	0.246094	0.006757
178.3	91.02	163.47	0.164063	0.007760
212.0	98.05	194.40	0.070313	0.008122
252.1	99.80	231.19	0.017578	0.008198
299.8	100.00	274.93	0.001953	0.008205

Equation (5.3) is used whenever a number count is taken and yields a specific surface if the surface-volume shape factor is known and vice-versa. If α_{sv} is assumed equal to 6 (this is assuming spherical particles), a specific surface is obtained. For microscope counting, for example, this is written as $S_{v,a}$ the volume-specific surface by microscopy since d_a is the particle projected area diameter. Similar arguments apply to equations (5.4) and (5.5).

Thus, the volume specific surface by sieving $S_{v,A} = 6 \times 0.008205 = 0.04923$ m^2 m^{-3}.

For a powder of density 2500 kg m^{-3}, the mass specific surface by sieving is :

$$S_{w,A} = 0.0197 \text{ m}^2 \text{ g}^{-1}.$$

Alternatively, if the specific surface is known, the surface-volume shape coefficient by sieving $\alpha_{sv,A} = S_v/0.008205$.

The surface-volume mean diameter of the distribution is defined as:

$$d_{sv} = \left(\sum \frac{q_r}{d_{A,r}} \right)^{-1} = 122 \ \mu m \tag{5.6}$$

as compared with the mass-median diameter for this distribution, of 126 μm.

5.3 Turbidity methods of surface area determination

When a light beam passes through a suspension the emergent intensity I is related to the intensity for a beam passing through the suspending liquid I_0 by the Beer–Lambert Law:

$$I = I_0 \exp(-\tau L) \tag{5.7}$$

where τ, the turbidity, is related to the projected area of the particles perpendicular to the direction of the light beam.

$$\tau = c \sum_{r=0}^{\max} k_r K_r \alpha_{sr} n_r d_r^2$$

where k_r is the shape coefficient relating particle projected area to particle diameter ($k_r = \pi/4$ for spherical particles) and K_r is the extinction coefficient.

Equation (5.7) can be written in terms of the optical density $D = \log(I_0 / I)$:

$$D = (\log_{10} e)\tau L$$

so that, the amount of light cut off by particles in a narrow size range centered on d_r is:

$$\Delta D_r = (\log_{10} e) c L k_r K_r n_r d_r^2$$

The laws of geometric optics break down as particle size approaches the wavelength of light and a correction term has to be included to compensate for this breakdown. This is in the form of an extinction coefficient K_r defined as the ratio of the amount of light cut off by the particle to that which would be cut off if the laws of geometric optics held.

If there are n_r particles of diameter d_r in 1 gram of powder ($W = 1$), the volume-specific surface is given by:

$$S_w = \sum_{r=0}^{\max} \alpha_{s,r} n_r d_r^2$$

where $\alpha_{s,r}$ is the surface shape coefficient (this relates particle diameter with surface area and equals π for an assembly of spherical particles $S = \pi d^2$) and:

$$S_w = \frac{\alpha_s}{kcL(\log_{10} e)} \sum_{r=0}^{\max} \frac{\Delta D_r}{K_r} \qquad (5.8)$$

assuming α_s and k_r are constants for the size range under consideration, i.e. particle shape does not change with size over the limited size range being examined.

According to Cauchy, for non-re-entrant particles the ratio of the surface and projected area shape coefficients is constant and equal to 4 [4]. For re-entrant particles the surface area obtained by making this assumption is the envelope surface area. Applying Cauchy's theorem simplifies equation (5.8) to:

$$S_w = \frac{9.2}{cL} \frac{D_m}{K_m} \qquad (5.9)$$

D_m is the maximum optical density and K_m is the average optical density over the size range under examination. If K_m is assumed equal to unity, equation (5.9) gives the mass specific surface by photo-extinction with no correction for the breakdown in the laws of geometric optics.

5.4 Adsorption from solution

The accumulation of one molecular species at the interface between a solid and a solution is governed by complex phenomena. The molecules may accumulate at the interface as a result of either chemical bonding or weak physical attractive forces. With physical adsorption, the molecules are easily removed by merely lowering the equilibrium concentration of the solution, whereas in chemisorption the molecules are more strongly attached.

Molecules are adsorbed at the interface by interaction of the unsatisfied field forces of the surface atoms of the solid with the force fields of the molecules striking the surface. In this way the free energy of the solid surface is diminished.

In adsorption from solution, solute and solvent molecules compete in the adsorption, and total coverage is difficult to compute since only the effect of the removal of solute molecules from solution can be determined.

The situation is further complicated by the large and complex shape of some solute molecules and the uncertainty in determining their orientation at the interface.

With some systems a saturation value is reached, resulting in a Langmuir type isotherm, making it possible to determine the monolayer capacity and eventually the specific surface. Giles *et al.* [5,6] examined the shape of a large number of published isotherms and divided them into four main types. They conclude that only a limited number of adsorbates can be used for surface area determination.

The usual experimental procedure is to add a known mass of powder to a solution of known concentration, which is maintained at a constant temperature, and agitate until equilibrium is attained. This may take a few hours or even a few days. Samples of the supernatant are then withdrawn and analyzed. Concentration measurements may be made using a spectrophotometer or a refractometer.

A more elegant approach is to circulate the solution through a bed of adsorbent and monitor the concentration continuously by passage through a *uv* cell [7, p. 21]

5.4.1 Molecular orientation at the solid–liquid interface

The idea of molecular orientation at an interface was conceived by Benjamin Franklin who, in 1765, spread olive oil on a water surface and estimated the thickness of the resulting film as one ten-millionth of an inch. Lord Rayleigh [8] in England and Miss Pockels [9] in Germany established that the film was only one molecule thick. Langmuir introduced novel experimental methods which resulted in new conceptions regarding these films.

Instead of working with oils, Langmuir used pure substances and measured the outward pressure exerted by the films directly using a floating barrier with a device to measure the force on it. The clearest results were obtained with long chain fatty acids and alcohols. Langmuir found that, as the surface on which the film was spread was reduced, no pressure developed until the area per molecule had been reduced to approximately 0.22 nm^2 at which point the pressure increased rapidly with further decrease in area. One of the most striking facts illustrated by Langmuir's work is that the area is independent of the number of carbon atoms in the molecules. This indicates that the molecules are oriented vertically to the surface of the liquid in the same manner in all the films regardless of the chain length. According to Adam [10] each molecule occupies an area of 0.205 nm^2 regardless of chain length. The orientation is such that the hydrophilic polar heads are in contact with the interface and the hydrophobic hydrocarbon chains are in the air.

Some investigators suggest that the effective area occupied by fatty acid molecules at solid–liquid interfaces is the same as that occupied by these molecules in films on water. The quoted areas for stearic acid, for example, ranges from 0.205 nm^2 to 0.251 nm^2, the former being the area of closest packing for ellipses and the latter the area for free rotation [11,12].

However, a greater variation than this is expected for an immobile interface since the adsorbate is not constrained to take up any definite orientation. In adsorption on carbon blacks Kipling and Wright [13] suggest that stearic acid is adsorbed with the hydrocarbon chain parallel to the surface, the effective area of each stearic acid molecule being calculated as 1.14 nm^2. Kipling and Wright [14] also suggest that this is true for other acids in homologous series, and adsorption of these acids on non-polar adsorbents indicates that the major axis of the hydrocarbon chain is parallel to the surface. This value was adopted by Roe [15] who applied the multilayer theory of adsorption to stearic and other aliphatic acids dissolved in cyclohexane at a volume concentration of 0.01235.

McBain and Dunn's [16] results for adsorption of cetyl alcohol on magnesium oxide are best interpreted in terms of orientation parallel to the surface. Smith and Hurley [17] determined the surface occupied by fatty acid molecules on to carbon black as 0.205 nm^2 which suggests a perpendicular orientation. Ward [18] suggested a coiling into a hemispherical shape and Allen and Patel [19,20] found that the area occupied by long-chain fatty acid molecules increased from 0.192 nm^2 to 0.702 nm^2 as the chain length increased irrespective of the adsorbent, while for long-chain alcohols the increase was from 0.201 nm^2 to 0.605 nm^2 [21] These values were interpreted in terms of coiling of the chains.

In early work on oxides [22,23], it was suggested that oleic acid and butyric acid adopted a perpendicular orientation on titania, as did stearic acid on aluminum hydroxide [24]. In these experiments it was not clear whether adsorption was physical or chemical in nature. This now seems an important distinction to draw, especially with basic solids. In chemisorption the orientation of the solute molecules generally presents no problem, as the functional group determines the point of attachment. Thus the long chain fatty acids are attached to the surface by the carboxyl group, –COOH, with the hydrocarbon chain perpendicular to the surface.

5.4.2 Polarity of organic liquids and adsorbents

Generally the organic liquids and solid adsorbents are classified according to their polarity. Polar molecules are defined as uncharged molecules in which the centers of gravity of positive and negative charges do not coincide, so that the molecules show dipole moments. The larger the dipole moment, the more polar the molecule. The term polar group is applied to a portion of the molecule with polar characteristics, such as –OH, –COOH, –COONa, –COOR and similar groups. Non-polar molecules have an equal number of positive and negative charges with coinciding centers of gravity. The dipole moment is zero for non-polar molecules. The term non-polar may be applied to a portion of a large molecule with non-polar characteristics such as benzene, *n*-heptane, hexane and other hydrocarbons.

The general rule is that a polar adsorbate will tend to prefer that phase which is the more polar, i.e. it will be strongly adsorbed by a polar adsorbent from a non-polar solution. Similarly, a non-polar adsorbate will be strongly adsorbed on a non-polar adsorbent from a polar solution.

Freundlich [25] found that the order of adsorption of normal fatty acids from aqueous solution on to a blood charcoal was formic, acetic, propionic, butyric in increasing order. The same order for homologous series of fatty acids, formic through caproic, from water on to charcoal was reported by Linnar and Gortner [26]. These results agree with Traube's rule [27]. Holmes and McKelvey [28] made a logical extension of Freundlich's statement by noting that the situation was a relative one and that a reverse order should apply if a polar adsorbate and a non-polar adsorbate were used. They indeed did observe a reverse sequence for fatty acid adsorbed on to silica gel from a toluene solution.

Langmuir [29] gave an instructive interpretation to this rule. The work to transfer one mole of solute from solution to surface is [30, p. 95]:

$$W = RT \ln\left(\frac{C_s}{C}\right) = RT \ln\left(\frac{\Gamma}{\tau C}\right) \tag{5.10}$$

where C_s is the surface concentration, Γ is the moles of solute adsorbed per unit area and τ is the film thickness. For solutes of chain length n and $(n-1)$ the difference in work is:

$$W_n - W_{n-1} = RT \ln\left(\frac{\Gamma_n}{\Gamma_{n-1}} \frac{C_n}{C_{n-1}}\right) \tag{5.11}$$

Traube found that for each additional CH_2 group the concentration required to give a certain surface tension was reduced by a factor of 3, i.e. if $C_{n-1} = 3C_n$ then $\gamma_n = \gamma_{n-1}$ and:

$$W_n - W_{n-1} = RT \ln 3$$

$$W_n - W_{n-1} = 2.68 \text{ kJ mol}^{-1} \text{ at } 20°C$$

This may be regarded as the work required to bring one group from the body of the solution to the surface. Adamson assumed this to imply that the chains were lying flat on the surface but suggested that this was undoubtedly an over simplification.

Harkins and Dahlstrom [31] have shown that the oxides of titanium, tin and zinc act like water in attracting polar rather than non-polar groups. Thus in oils any –COOH, –OH, –CN and other similar groups orient towards the oxide powder and the hydrocarbon groups orient themselves towards the oil.

5.4.3 Drying of organic liquids and adsorbents

In adsorption by solids from liquid phases, substances in solution at very low concentrations are often preferentially adsorbed; the presence of trace quantities of water and other impurities in the solution may therefore have an effect on adsorption. Harkins and Dahlstrom [31], for example, reported that extremely small quantities of water in benzene increased the energy of immersion of oxides to about three times the value obtained with pure benzene. In order to obtain liquids of sufficient purity it may be necessary to fractionate and then store over metallic sodium or other drying agent such as silica gel, calcium sulphate, alumina and so on.

Most solid adsorbents are capable of adsorbing water vapor from the atmosphere and should therefore be dried. This is usually done by heating for 2 or 3h at a temperature between 120°C and 130°C.

Some workers [32,33] claim that this temperature is sometimes not high enough to drive away vapors previously adsorbed on to the solids. If a higher temperature is used care has to be taken that the surface of the solid is not altered in any way e.g. that sintering or surface texture are not changed.

Some experimenters consider that adsorbents should be degassed before use and then introduced to the solution in the absence of air. Others claim that such outgassing treatment does not affect the extent of adsorption. Thus it was reported by Greenhill [34] and by Russell and Cochran [35] that adsorption was the same on metals, metal oxides and non-porous alumina, whether the samples were outgassed or not prior to exposure to the solutions; gases apparently being displaced by the liquid phase [36]. No systematic effect was found for adsorption on charcoal from mixtures of carbon tetrachloride and methanol [37]. Hirst and Lancaster [38] examined the effect of very small quantities of water on the interaction of stearic acid with finely divided solids. For adsorbates such as TiO_2, SiO_2, TiC and SiC, the presence of water reduced the amount of acid adsorbed to a monolayer, and with reactive material such as Cu, Cu_2O, CuO, Zn and ZnO, water was found to initiate chemical reaction.

A useful method of purification of solvents has been given by Weissberger and associates [39].

5.5 Theory for adsorption from solution

Liquid phase adsorption methods depend on the establishment of an equilibrium between adsorbed and unadsorbed solute molecules. Adsorption of solute on to the surface of a solid continues until it reaches a saturation point giving a clear plateau in the isotherm. As the isotherm usually tends towards a limiting value, the limit has often been taken to correspond to the covering of the surface with a monolayer of solute. The equation derived for monolayer coverage is:

$$\frac{x_1 x_2}{\Gamma_1^{(n)}} = \frac{1}{K x_m} + \frac{K-1}{K} \frac{x_1}{x_m} \tag{5.12}$$

Where $\Gamma_1^{(n)}$ is the Gibbs isotherm value x_1, x_2 are the mole fractions of the two components of a completely miscible solution and K is a constant.

For K much greater than unity and for low concentrations of component one, this reduces to Langmuir's equation. Alternatively, the Langmuir equation, replacing p with concentration of solution c, has been used to determine the limiting value [19,20].

$$\frac{c}{x} = \frac{1}{K x_m} + \frac{c}{x_m} \tag{5.13}$$

where $S_w = \dfrac{N \sigma x_m}{M_v}$

and
x = amount of solute adsorbed per gram of adsorbent;
x_m = monolayer capacity;
K = constant;
σ = area occupied per molecule;
M_v = molar volume.

Thus a plot of c/x versus c should give a straight line of slope $1/x_m$ and intercept $1/K x_m$.

Three things are required for the determination of specific surface, namely:

1). The area (σ) occupied by one molecule of the adsorbate in a close packed film on the surface of the adsorbent.

2). The point on the isotherm where a complete monolayer has been formed must be clearly located.

3). It must be possible to compensate for competitive adsorption of solvent molecules on the adsorbent surface.

5.6 Methods for determining the amount of solute adsorbed

In almost all studies of adsorption from solution it is necessary to measure the concentration of the solution before and after adsorption. A variety of analytical methods may be used to measure such changes in concentration, including the Langmuir trough, gravimetric, colorimetric, titrimetric, interferometry and precolumn methods.

5.6.1 Langmuir trough

This technique [40] can be useful where the adsorptive can be spread on an aqueous substrate to give a coherent film. The area occupied by the absorptive film, after evaporation of the solvent, is proportional to the amount of solute present. This method has been applied successfully by Hutchinson [41], Gregg [42] and Greenhill [43] to analysis of solutions of long-chain fatty acids and by Crisp [44] to alcohols and phenols in organic solvents such as benzene. Equal volumes of solution were spread on an aqueous substrate before and after adsorption.

5.6.2 Gravimetric methods

If an involatile solute is dissolved in a volatile solvent, analysis can be effected by evaporating off the solvent from a sample of known weight and weighing the residue. This simple technique was adopted by Smith and Fuzek [45] and thereafter widely used. In their procedure an estimated 0.5 to 1 g of adsorbent was placed in a glass sorption tube to which a vacuum source could be attached. 40 mL of fatty acid solution (0.15 g) in benzene was introduced and the tube was stoppered and shaken for a definite period of time. The tube was then centrifuged in order to settle the adsorbent and 5 mL of supernatant was withdrawn. This liquid was delivered into a tared container which was placed in an oven at a temperature just below the boiling point for the solvent. Evaporation of the solvent was speeded up by means of a slow stream of filtered air, then the amount of remaining fatty acid was determined by weighing. Blank runs established the dependability of this analytical procedure. The adsorption tube was re-stoppered, shaken again for a definite time, and some of the supernatant removed as before. At the end of the experiment the adsorbate was filtered out, dried and weighed under CO_2 and the increased weight of the adsorbent determined.

5.6.3 Volumetric method

Many standard procedures are available for studying adsorption by volumetric or titrimetric methods. Adsorption of fatty acid [17] has frequently been determined by titration with aqueous alkali, even if the fatty acid was previously dissolved in organic solvent. Extraction of fatty acid from the solvent caused no problems, especially if warm ethyl alcohol is added [13], but the validity of the method needs to be checked by titration of a known sample each time.

Conductimetry [46] and potentiometry [47] titrations have been used as an alternative to those carried out with a colored indicator.

5.6.4 Rayleigh interferometer

This instrument is used to measure the difference in refractive index or in optical path length between two liquids by a 'null' method. The measurement is converted into a difference in composition by means of a calibration curve. The use of the instrument is restricted to systems with a small difference in refractive index, otherwise a large number of standard mixtures have to be made up for calibration purposes.

Bartell and Sloan [48] and Ewing and Rhoda [49] have made successful use of the interferometer in measuring the change in concentration due to adsorption from non-aqueous solutions. Further details of the instrument are given by Candler [50].

5.6.5 Precolumn method

This was suggested by Groszek [51] for measuring the amount of solute adsorbed on to a solid surface using a flow microcalorimeter and is described in detail later.

5.7 Experimental procedures for adsorption from solution

5.7.1 Non-electrolytes

This is usually considered to be essentially monolayer adsorption with competition between solvent and solute. The non-electrolytes that have been studied are mainly fatty acids, aromatic acids, esters and other single functionless group compounds plus a great variety of more complex species such as porphyrins, bile pigments, carotenoids, lipoids and dyestuffs.

5.7.2 Fatty acids

When fatty acid molecules are closely packed on the surface of distilled water, each molecule occupies an area of 0.205 nm^2 irrespective of the length of the hydrocarbon chain. This property was used by Harkins and Gans [22] for the determination of the surface area of titanium dioxide using oleic acid, with results in general agreement with microscopy. Since then the method has been used extensively and a detailed review is to be found in Orr and Dallavalle [52]

Smith and Fuzek [47] used the gravimetric method described earlier. They also used a procedure in which 20 mL of solvent containing 0.2 to 0.4 g of fatty acid were placed in a tube and mixed for 24h. After this time the tube was centrifuged, 10 mL of the liquid was withdrawn and 10 mL of pure solvent added. The procedure was then repeated. The fatty acid content of the withdrawn samples was determined as before.

Gregg used the same solvent [53] but determined the amount adsorbed using a surface tension balance. Smith and Hurley [17] recommended the use of cyclohexane as solvent, and stated that with some solvents multilayer adsorption takes place. Hirst and Lancaster [54], instead of adding more fatty acid to the solvent, increased c/c_s by lowering the temperature of the solution.

The specific surface area determined by liquid phase adsorption is usually low due to solvent adsorption. It is thus preferable that determinations be carried out with a variety of solvents and comparisons made between them.

5.7.3 Polymers

Adsorption isotherms of linear polymer molecules are found to be of the Langmuir type [55,56]. Many workers assume that the molecules are adsorbed in the shape of a random coil [57,58] and have developed equations to give the area occupied by a molecule. Some workers assume a modified Langmuir equation to be necessary since polymers may occupy more than one site [59]. Others adopt a more empirical approach [60]. An estimate of the inner and outer surface areas of porous solids has also been obtained by using a set of polystyrene fractions having narrow ranges of molecular weights [61].

5.8 Adsorption of dyes

Dyes have been used by many investigators for specific surface evaluation but the procedure has not been widely accepted because of the inconsistency of the reported results both between different dyes and with other methods [62,63]. Giles *et al.*[64] attribute these inconsistencies mainly to injudicious choice of dyes and incomplete understanding of the adsorption process.

Extensive studies of the adsorption of non-ionic [65] and cationic [66,67] dyes on alumina and inorganic surfaces were carried out by Giles and his co-workers [68,69] who consider that these dyes are adsorbed by an ion exchange mechanism, largely as a monolayer of positively charged micelles or aggregates. The presence of aggregates on the substrate is inferred from the nature of the isotherm and its change with temperature.

Anionic and non-ionic compounds with free hydroxyl groups are adsorbed on silica mainly as a monolayer of single molecules or ions with the hydroxyl group bonded to the silica surface, but micelle adsorption may also occur. Leuco vat dyes [70] are adsorbed in aggregate form, but the evidence is not wholly unambiguous since it is based on the spectral characteristics of the adsorbed dye on the substrate and also its fluorescence characteristics [71]. Non-reactive anionic phthalocyanine blue is adsorbed in aggregate form to give an appreciable percentage of dimers [71].

Giles *et al.*[70] discussed the orientation of molecules at graphite surfaces and concluded that non-ionic azo dye is adsorbed flat and a complete monolayer is formed from aliphatic solvents, but not from benzene due to competition from the benzene molecules. With monosulphonates a condensed monolayer is formed; the molecules probably standing perpendicular to the surface, the sulphonated groups in the water and the unsulphonated ends at the graphite surface. Disulphonate molecules are oriented edge on or end on and trisulphonate molecules lie flat.

On anodic alumina films, sulphonated dyes can be adsorbed both by covalent bonds between the sulphonate groups and alumina atoms and also by electrostatic attraction of ionic dye micelles by the positively charged surface [64].

The experimental procedure is to gently tumble 0.05 to 0.50 g of sample in 10 mL in a centrifuge tube containing aqueous solution of dye at room temperature; 10 to 30 min is sufficient for non-porous solids, but 12 to 48h may be required for porous powders. The tubes are then centrifuged and the solutions analyzed spectrophotometrically. With porous powders, a rate curve develops; extrapolating this back to zero gives the surface concentration of dye; the saturated value representing total coverage. The isotherms usually have a long plateau and this value is accepted as monolayer coverage; this feature makes the method attractive as a one-point technique.

If Y_m (in mmol g^{-1}) is the amount of dye adsorbed at monolayer coverage per gram of adsorbent, σ the flat molecular area of the dye and N is the Avogadro constant, the weight specific surface is given by:

$$S_w = \frac{Y_m N \sigma}{X} \tag{5.14}$$

where X is the coverage factor which is equal to the number of dye ions in a micelle.

For methylene blue Giles [72] use $\sigma = 1.2$ nm^2 and $X = 2$. They found it preferable to buffer the dye solution at pH 9.2, by using buffer tablets, to reduce competition between the H$^+$ and the dye cations. In a later paper they advise against tests being made from solutions outside the pH range 5–8 [73].

Padday [74] advises the avoidance of dyes which may be adsorbed in a dual manner and, indeed, any dyes which may form covalent bonds with the surface, since they may be liable to selective adsorption on particular sites. They recommend the use of methylene blue BP, brilliant basic red B, crystal violet BP, victoria pure lake blue BO, orange II or solway ultra-blue. The two BP dyes can be used as bought, the rest need some pretreatment.

The surface areas of fibers was determined by Mesderfer *et al.* [75] with Direct Red 23 cation in cold water. Parallel tests were carried out on carbon black of known specific surface and the resulting isotherms compared.

A review of dye adsorption has been presented by Padday [74], who concentrated particularly on the use of cyanine dyes for the surface area measurement of silver bromide and silver iodide.

5.9 Adsorption of electrolytes

There are several variations to this technique. The negative adsorption method is based on the exclusion of co-ions from the electrical double layer surrounding charged particles [76].

The ion exchange method is based on replacing of loosely held ions by others of the same sign [30, p. 593]. A large amount of work has been done on the adsorption of electrolytes by ionic crystals and the adsorption of ions from solution on to metals. Since the adsorption tends to be very small and the measurements rather tedious, these are not suitable as routine methods.

5.10 Deposition of silver

The surface area of paper pulp fibers has been determined by the deposition of a continuous film of metallic silver on the surface by use of the reducing properties of cellulose [77]. The amount of metal deposited was determined by its ability to decompose hydrogen peroxide catalytically. A standard surface, e.g. regenerated cellulose film, is required for calibration.

5.11 Adsorption of *p*-nitrophenol

Giles *et al.* [5,6] have investigated the adsorption of *p*-nitrophenol (PNP) for surface area evaluation. It is normally used in aqueous solution but can be used in an organic solvent [72,73]. The method is recommended as being suitable for a wide variety of solids, both porous and non-porous, provided they either form a hydrogen bond with PNP or have aromatic nuclei. With porous solids the results may or may not reveal a lower accessible surface than nitrogen adsorption, depending on the pore size distribution. With porous solids a rapid estimate of the relative proportions of large and small pores can be made, together with a measure of external surface. PNP has also been used, by a modified technique, to measure the external surface, as opposed to the much greater inner surface, of fibers [78].

Normally PNP is adsorbed flatwise with an effective molecular area of 0.525 nm^2. In some cases, on polar inorganic solids, it is adsorbed end-on with an effective area of 0.25 nm^2. The mode of adsorption is indicated by the type of isotherm generated, an S-type curve indicating end-on adsorption.

The experimental procedure is similar to that used for dye adsorption [79]. Since PNP does not adsorb on cellulose it can also be used for the measurement of specific surface of colorants in cellulose [80].

The use of PNP adsorption from aqueous solutions on to porous adsorbents such as silica gel or carbons has been queried by Sandle [cit. 81] who states that competition of water molecules for the surface will yield erroneous results. Giles agrees that complete coverage is not always obtained with certain acidic solids and recommends the use of other solvents.

Padday [74] states that the accuracy of PNP is suspect since many isotherms show no clearly defined plateau.

5.12 Chemisorption

Physisorption is a superficial phenomenon similar to the condensation of a vapor on to a surface and the energy involved is the same order as the heat of condensation of the adsorbed gas.

Chemisorption involves the sharing of electrons between the adsorbate and adsorbent and the energy involved is of the same order as chemical reaction energies.

Although an upper energy level for physisorption of 40 KJ mol^{-1} is sometimes quoted, there is considerable overlap between the two phenomena.

Chemisorption is limited to a monolayer and the shape of the isotherm is predicted by the Langmuir equation.

The area occupied by each adsorbate molecule depends upon the availability of active sites; because of this it is particularly useful for determining the active surface in a multicomponent system.

5.12.1 Hydrogen

Spenadel and Boudart [82] determined the area of platinum black by the physical adsorption of argon and compared this to the area determined by chemisorption of hydrogen. Assuming that each atom of hydrogen is attached to one atom of platinum, then the area occupied by each atom is 0.89 nm^2. The subsequent calculation generates a similar surface area for the two techniques. Adams calculated the average area per chemisorbed molecule of hydrogen on platinum to be 224 nm^2 to give a site area of 112 nm^2 per atom of platinum [83]. The difference can be attributed to differences in the particle size of the platinum crystallites [84].

5.12.2 Oxygen

The chemisorption of oxygen on metals is more complicated than the chemisorption of hydrogen since this is normally the first step in oxidation, leading to the formation of a thick oxide layer on the surface.

5.12.3 Carbon monoxide

The chemisorption of carbon monoxide is an established method for determining the surface area of dispersal metals, particularly in supported catalysts. The average area occupied by each molecule depends on whether attachment is on one or two sites, a state that can vary from metal to metal and with surface coverage [85]. The quantity of chemisorbed gases is commonly measured by volumetric methods with apparatus similar to that used for static BET gas adsorption measurements.

5.13 Other systems

Adnadevic and Vucelic [86] compared the sorption of polar molecules with the BET gas adsorption technique for the surface area of microporous solids. He examined activated charcoal, zeolites and silica clay using MeOH, EtOH, CO_2 and C_6H_6.

Larionov [87] reviewed specific area determination by adsorption from solution. Veselov and Galenko [88] used a 1:1 volume ratio of C_6H_6–heptane solution and found good agreement with BET. For the adsorption of nitrobenzene, *p*-chloroaniline, and *p*-nitroaniline on active carbons it was found that the Langmuir and D–R isotherm equations were obeyed [89]. Pore size distribution determination were also carried out using the *t*-method with C_6H_6 as adsorbate.

Gata [90] used ethylene glycol to examine clay minerals. Mazden [91] determined total surface (internal and external) by removal of a previously adsorbed two-layer glycerol complex by assay of the weight loss on heating. Kulshreshtha *et al.* [92] estimated fiber porosity from the glycerol retention value.

The temperature change and dependence on time, during dissolution rate studies, have also provided information on particle size distribution and surface [93]. Ruzek and Zbuzek [94] determined surface area from dissolution rates and compared their results with electron microscope data.

A rapid method of particle size evaluation of carbon black by iodine number has also been described [95]. Chetty and Naidu [96] determined the surface area of sulfides by the isotope exchange method using tracer [203]Hg for CdS and HgS and [124]Sb for Sb_2S_6.

Corrin *et al.* [97] determined the adsorption isotherm of sodium dodecyl sulfate and potassium myristate on ash-free graphite. Calculations based on two extreme assumptions concerning the concentration of solvent in the surface region yielded specific surface values which differed by less than the experimental error. The isotherm of sodium dodecyl sulfate exhibited a discontinuity at the critical concentration for micelle formation. The adsorption of sodium dodecyl sulfate on polystyrene was also measured. The minimum area per molecule was 5.1 nm^2 for the sulfate and 3.66 nm^2 for the myristate. The experimental procedures were described in detail.

5.14 Theory for heat of adsorption from a liquid phase

5.14.1 Surface free energy of a fluid

A fluid has a surface energy only when it exists in a sufficiently condensed state. Because of the uniform energy distribution in a gas, no difference exists between an internal molecule in the center of a gas volume and a molecule located near a wall; therefore a gas has no surface energy.

The theory for the forces acting between molecules was put forward by Lennard-Jones and Devonshire [98]. The liquid molecule is assumed to be located in a cage formed by the neighboring molecules, and it is constantly under the influence of their fields yet being sufficiently free to execute translatory and rotary movements.

Each molecule in a liquid volume is surrounded by molecules on all sides, and hence is subjected to attractive forces acting in all directions. Generally speaking, a uniform attraction in all directions is exerted by every molecule for a period of time which is relatively long compared with periods of vibration.

Very different conditions obtain at the surface. The molecules are attracted back towards the liquid and also from all sides by their neighbors, yet no attraction acts outwards to compensate for the attraction towards the center. Each surface molecule is subjected to a powerful attraction towards the center acting, for reasons of symmetry, in a direction normal to the surface.

The work required to increase the area of the surface by an infinitesimal amount dA, at constant temperature, pressure and composition is done against a tension γ, generally known as the surface tension which can be defined from the point of view of energy involved [99,100]. The free energy change dF is equal to the reversible work done:

$$dF = \gamma \, dA$$

$$\gamma = \left(\frac{dF}{dA} \right)_{T,P,n} = F_s \qquad (5.15)$$

where F_s is the surface free energy per unit area.

Surface tension and surface free energy are, in effect, two different aspects of the same matter. In the SI system, the number which, in newtons per meter, indicates the surface tension will, in joules per meter^2, express the surface free energy. The two equations above express a fundamental relationship in surface chemistry.

5.14.2 Surface entropy and energy

The entropy of a system at constant pressure, surface area and composition is:

$$-S = \left(\frac{\mathrm{d}F}{\mathrm{d}T}\right)_{P,A,n} \tag{5.16}$$

For a pure liquid the surface entropy per unit area S is:

$$-S = \frac{\mathrm{d}\gamma}{\mathrm{d}T} \tag{5.17}$$

The total energy per unit area for a pure liquid E is:

$$E = F_s - T\frac{\mathrm{d}\gamma}{\mathrm{d}T} \tag{5.16}$$

It is the work which must be done in order to remove from the bulk of the liquid and bring to the surface a sufficient number of molecules to form a surface unit. Conversely, energy is liberated when a surface disappears due to the return of molecules from the surface to the center of the liquid. The total surface energy of a pure liquid is generally larger than the surface free energy.

5.14.3 Heat of immersion

The theory for heat of immersion is due to Harkins [101] and Bangham and Razouk [102–104] .

When a clean solid surface is immersed in or wetted by a liquid, it leads to the disappearance of the solid surface and the formation of a solid–liquid interface. As a result of the disappearance of the solid surface, the total energy of the solid surface is liberated. The formation of the solid–liquid interface leads to an adsorption of energy equal to the energy of the interface. Thus, from thermodynamic considerations, the heat of immersion (E_{imm}) is equal to the surface energy of the solid–gas (or vapor) interface (E_{SV}) minus the interfacial energy between the liquid and the solid (E_{SL}), so:

$$E_{imm} = (E_{SV} - E_{SL})S$$

$$-q_0 = \frac{E_{imm}}{W} = S_w(E_{SV} - E_{SL}) \tag{5.17}$$

where $-q_0$ is the heat evolved per gram of solid, S_w is the weight specific surface of the solid and E the energies per unit area. By analogy with equation (5.16), the surface energies when solid-liquid and solid-vapor are in contact are:

$$E_{SL} = \gamma_{SL} - T\frac{d\gamma_{SL}}{dT}$$

$$E_{SV} = \gamma_{SV} - T\frac{d\gamma_{SV}}{dT}$$

which, on combination give:

$$(E_{SV} - E_{SL}) = (\gamma_{SV} - \gamma_{SL})T\frac{d(\gamma_{SV} - \gamma_{SL})}{dT} \tag{5.18}$$

This may be simplified by the use of the adhesion tension relationship of Young and Dupre [cit. 30]

$$\gamma_{SV} = \gamma_{SL} - \gamma_{VL}\cos(\theta) \tag{5.19}$$

where θ is the angle of contact between the solid and the liquid. Inserting equation (5.19) in (5.18) and simplifying gives:

$$(E_{SV} - E_{SL}) = \gamma_{LV} - T\frac{d\gamma_{LV}}{dT}\cos(\theta) \tag{5.20}$$

Finally, substituting equation (5.20) into (5.17) and considering that a liquid which wets a solid has zero contact angle:

$$-q_0 = S_w\left(\gamma_{LV} - T\frac{d\gamma_{LV}}{dT}\right) \tag{5.21}$$

5.15 Static calorimetry

The calorimetric method may be used in two ways; immersion of the bare outgassed solid in pure liquid, and immersion of the solid precoated with the vapor phase. The former approach is not widely used because of problems due to the state of the surface (impurities and defects), although the heats of immersion per unit area of a number of liquid/solid systems are known. In order to reduce surface effects Chessick *et al.* [105] and Taylor [106] used liquid nitrogen; the method is limited since it is not applicable to microporous solids and a surface area in excess of 150 m^2 is required.

If the solid is first equilibrated with saturated vapor, then immersed in pure liquid adsorbate, the solid–vapor interface is destroyed and the heat liberated should correspond to E_L, the surface energy of the pure liquid. The above assumption is made in what is termed the absolute method of Harkins and Jura (HJa) [107] who obtained a heat of immersion of 1.705 kJ kg^{-1} for titanium dioxide which, when divided by the surface energy of the adsorbent, water (11.8 kJ kg^{-1}) gave a surface area of 14.4 m^2 g^{-1} in agreement with the BET value. For a comprehensive bibliography and description of the calorimeter used, readers are referred to Adamson [30]. The validity of the HJa method may be questioned because exposure to a saturating vapor causes capillary condensation which reduces the available surface. A correction is also required for the thickness of the adsorbed film.

The same technique was used by Clint *et al.* [108] for the determination of the surface of carbon black by the adsorption of *n*-alkanes. Equation (5.21) was used with the following correction for small particles where the thickness of the adsorbed layer *t* was not negligible in comparison with the particle radius *r:*

$$S'_w = \left(\frac{r}{r+t}\right)^2 S_w \qquad\qquad (5.22)$$

For low surface areas this method gave reasonable agreement with other techniques, but the measured surface areas for particles with high surface areas were too low. The method is essentially comparative since the entropy is obtained using a reference sample. The method was considered to be unsuitable for powders having a specific surface smaller than 20 m^2 g^{-1}.

Partyka *et al.* [109]examined seven non-porous adsorbents with water and amyl alcohol as adsorbates and found agreement with BET for five. The heat of immersion values indicated that the numbers of adsorbed layers were between 1 and 2 as opposed to the 5 to 7 layers (1.5 to 2 nm) found by Harkins and Jura. They later expanded the work using mainly water but also long polar–non–polar molecules such as pentanol and butanol and the non–polar molecule decane [110]. They equated q_0 with enthalpy and showed that the curve of immersion enthalpy reached a plateau when the sample is equilibrated with 1.5 layers pre-adsorbed: from this coverage onwards the interface behaves like a bulk liquid surface. The surface areas using water were in good agreement with BET surface areas; there was a long range interaction with butanol which rendered it unsatisfactory; decane was satisfactory; pentanol compared well with water and deemed satisfactory for hydrophobic surfaces.

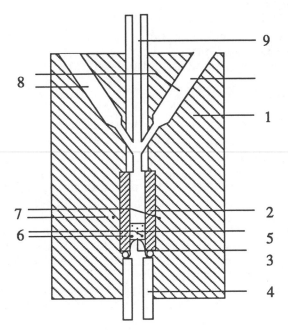

Fig. 5.1 Schematic diagram of the Microscal flow microcalorimeter.

A review of calorimetric methods of surface area determination has been presented by Zettlemoyer *et al.* [111–112], who also discuss some of the difficulties encountered in the measurement and interpretation of heats of immersion. They also developed a static microcalorimeter which was capable of determining surface areas as low as 1 m^2 g^{-1} [113].

5.16 Flow microcalorimetry

A variation of this technique is found with the flow microcalorimeter (Figure 5.1). The calorimeter consists of a metal block (1) surrounding a cylindrical cavity in which the PTFE calorimeter cell (2) is situated. The cell forms a continuation of the inlet tube for the carrier fluid and is joined to the outlet tube(4). The outlet tube is fitted with a 200 mesh stainless steel gauze (3) on which the powder bed (5) is placed. For analyses at ambient temperature, flow is normally through the central cavity but at elevated temperatures the fluids are fed through cavities (8). The heat generated is determined by measuring (6) and reference (7) thermisters in a Wheatstone bridge network.

5.16.1 Experimental procedures– liquids

There are three main experimental methods of studying the heat of adsorption at a solid–liquid interface by using the flow microcalorimeter.

1   pulse or injection adsorption;
2   equilibrium adsorption;
3   successive or incremental adsorption.

(a) Pulse adsorption

The powder bed is made up of a few tenths of a gram of powder through which the solvent flows at a flow rate of around 1 mL min^{-1}. In the pulse method a micrometer syringe is used to introduce small quantities of surface-active substances (e.g. 1–100 μg as 0.1 to 1% solutions). These are introduced into the stream of carrier liquid against the wall of the inlet tube below the point at which the liquid leaves the flow control capillary (9).

Any changes in the calorimeter cell are registered on a recording device. Normally, adsorption is accompanied by the evolution of heat in the bed which generates a positive pulse on the recorder. In the case of irreversible (chemisorption) change, the pen will return to the baseline, but if the change is reversible (physisorption) the pen will cross the baseline to describe a negative heat of desorption.

(b) Equilibrium adsorption

In this method two reservoirs are prepared, one of which contains the pure carrier liquid (solvent) and the second the solution of active agent. Initially a steady flow of carrier liquid is passed through the adsorbent bed. When the calorimeter comes to thermal equilibrium, giving a steady baseline on the recorder, the flow of carrier liquid is replaced by solution. Care needs to be taken that the two flows do not differ by more than 0.01 mL min^{-1}.

Adsorption of solute, which is accompanied by a heat and hence a temperature change, is measured by the thermisters which are connected via a Wheatstone network to the recorder. The result for an exothermic reaction (heat change typical of adsorption) is a positive pulse on the recorder which then returns to the base line when the adsorption of solute is complete. Desorption may be carried out by returning to a flow of the carrier liquid. The result for an endothermic reaction (heat change typical of desorption) is a negative pulse on the recorder.

For physical adsorption the area under the pulse is the same as the area under the adsorption peak. The rate of desorption, which controls the shape of the desorption pulse, depends on the relative strengths of adsorption of the solute and solvent. In some instances a long time elapses before all the solute molecules are removed, and this results in a desorption pulse having a long tail, making it difficult to decide when desorption is complete.

In chemical adsorption the heat generated is greater than for physical adsorption (greater than 4 kJ mole^{-1}). Further, the desorption peak is much smaller than the adsorption peak.

Adsorption–desorption of various solute concentrations may be studied on the same plug or different plugs of adsorbent. This way the form of the adsorption isotherm can be determined.

In practice it is found that using the same plug of adsorbent and different concentrations of solution is not a good method to adopt for the adsorption process due to the difficulty of determining when desorption is complete. It is therefore preferable to use a different plug for each point on the isotherm.

(c) Successive adsorption

In this method of using increasing concentrations of the solution for adsorption, with no desorption cycles, a series of heat effects occur (pulses), and these generally decrease for equal increments in solute concentration. When the adsorbent surface is completely saturated with the solute, further runs at increasing concentrations do not give any heat effects. Adding the pulse areas gives the same result as using the equilibrium adsorption technique. The method is however simpler, although errors in determining pulse areas are compounded. This method is therefore the preferred one.

5.16.2 Calibration

Calibration is effected by replacing the standard outlet tube with one containing a heating coil. With a powder bed in position and the carrier liquid flowing, known quantities of electrical energy are injected into the system and a calibration curve generated of areas under the pulses against heat injected.

5.16.3 Precolumn method

In constructing adsorption isotherms it is necessary to measure the amount of solute adsorbed at a range of concentrations. This is usually done by measuring the concentration of solute before and after adsorption. The following is a variation on that technique.

The adsorbent is placed in a precolumn constructed from a glass tube with the same internal diameter as the calorimeter cell. The carrier liquid is then percolated through the adsorbent in the precolumn before it enters the adsorbent in the calorimeter. To determine the amount of solute adsorbed from a given solution, the flow of carrier liquid is stopped at time t_1 and the flow of solution started. If the precolumn contained an inert solid, the solution would emerge at time t_0 and would then contact the adsorbent in the calorimeter, but when the precolumn contains an adsorbent the solute is retained, so that only solvent emerges at time t_0, the solution emerging at a later time t_2. The difference between t_2 and t_1 is a measure of the amount of solute adsorbed.

In the above method the critical points are the accurate determination of the time at which the solution emerges from the precolumn and an accurate knowledge of the flowrate of the solution through the precolumn. It is also necessary to ensure, for all adsorbates, that the flowrates are sufficiently low to permit full saturation of the adsorbent in the precolumn. If the flowrate of the solution is too high, solution may emerge from the precolumn before the adsorbent is fully saturated. In such cases the estimate for the amount of solute adsorbed would be lower than the equilibrium value.

However, if the retention time $(t = t_2-t_1)$ and flow rate are known the amount of solute adsorbed can be calculated from the equation:

$$x = \frac{tCf}{w} \qquad (5.23)$$

where x = amount of solute adsorbed (mg g^{-1})
  C = concentration of solute (mg mL^{-1})
  f = flow rate of solution (mL min^{-1})
  w = weight of adsorbent (g)

5.16.4 Experimental procedure – gases

A similar procedure can be carried out with gases. This technique has been used with carbon dioxide as the adsorbate and nitrogen as the carrier gas [114]. Adsorbents used were γ-alumina, Gasil and Degussa silica; experiments were performed at 22°C and the results compared with carbon dioxide adsorption at 22°C and –78°C.

The carbon dioxide isotherm at 22°C on γ-alumina was found to obey Sips' equation [115], which is valid for adsorption, at low pressure, on a non-uniform surface.

$$\frac{V}{V_m} = \theta = \left(\frac{p}{p+B}\right)^c \qquad (5.24)$$

The procedure to determine V_m is to select the value of B to give the best fit when log V is plotted against $p/(p+B)$.

The heat of adsorption of carbon dioxide on γ-alumina decreased with increasing coverage, thus substantiating that the surface was non-uniform. The surface of Gasil silica was found to be slightly heterogeneous and that of Degussa silica was found to be homogeneous. In later papers Allen and Burevski examined the adsorption of gases on microporous carbons [116] and the adsorption of sulfur dioxide on powdered adsorbents [117].

5.16.5 Applications to surface area determination

Using liquid flow microcalorimetry it is possible to determine the area occupied per adsorbate molecule and the energy of adsorption per molecule (actually the energy required to displace a solvent molecule by a solute molecule). Allen and Patel [19,20] investigated a range of long chain fatty acid and long chain alcohol adsorbates and obtained information on molecular orientation at solidsolution interfaces. Surface areas were calculated using the following form of equation (5.21):

$$S_w = K_0 q_m \qquad (5.25)$$

where q_m is the Langmuir monolayer value for heat of displacement of solvent molecules by solute molecules. For the adsorption of *n*-octoic acid from *n*-heptane, $K_0 = 16.7$ m^2 J^{-1}, and two thirds of the determined values agreed with BET values to within 10%. Full details of the experimental procedures and results are given in [118].

Burevski found that the energy of adsorption for gases varied with coverage [119]. The manner in which it varied depended on the system under examination, rendering the method unsuitable for surface area determination.

5.17 Density methods

The total voidage and porosity of a compound, i.e. tablet or compacted plug of material, can be determined if the density of the matrix material is known. It is necessary to determine the envelope volume occupied by the compound and this can be done using a pyknometer and a non-wetting liquid. If necessary the surface can be rendered non-wettable by a surface treatment [120, p. 30]. An error will still arise due to hydrostatic pressure forcing some of the liquid into the pores. This can be prevented by coating the compound with an impermeable coating of known or negligible density.

The total voidage may also be determined by saturating the sample with a wetting liquid after removing previously adsorbed vapor by vacuum evaporation [121].

The gas comparison pyknometer is used for skeletal density measurement and it is found that, when used with air, erroneous results are obtained and sometimes even negative volumes are found. This indicates that air is being adsorbed and the amount adsorbed can be determined by comparison with helium density measurements. As nitrogen and oxygen, at room temperature, obey Henry's law the measured adsorption on increasing the pressure from 1 to 2 atm. is practically equivalent to the air adsorption at 1 atm. The relevant equation is:

$$wS_w = CV_a$$

$$wS_w = C(V_{He} - V_{N_2})$$

$$wS_w = C(\frac{1}{\rho_{He}} - \frac{1}{\rho_{N_2}}) \qquad (5.26)$$

where

S_w	is the weight specific surface of the sample;
V_a	is the volume of air adsorbed as the pressure is increased from 1 atm to 2 atm;
w	is the weight of the powder;
V_{He}, V_{N_2}	are the indicated powder volumes using helium and nitrogen respectively;
ρ_{He}, ρ_{N_2}	are the indicated powder densities using helium and nitrogen respectively;
C	is a constant.

Tuul and Innes [122] reported a variation of C with substrate whereas Jäkel [123] stated that C was constant for a wide range of adsorbents.

References

1   BS 4359 Determination of the specific surface area of powders: Part 1: *Recommendations for gas adsorption–BET methods* (1984), *191*

2   BS 4359 (1982), Determination of the specific surface area of powders: Part 2: *Recommended air permeability methods* (), *191*

3   BS 4359 (1973), Determination of the specific surface area of powders: Part 3: *Calculation from the particle size distribution*, *191*

4   Cauchy, A. (1840), *C. R. Acad. Sci.*, Paris, **13**, 1060, *196*

5   Giles, C.H., MacEwan, T.H., Nakhwa, S.N. and Smith, D. (1960), *J. Chem. Soc.*, 3973, *196, 206*

6   Giles, C.H. and Nakhwa, S.N. (1962), *J. Appl. Chem.*, **12**, 266, *196, 206*

7   Sing, K.S.W. (1992), *Proc. Conf. Particle Size Analysis*, Loughborough, Anal. Div. Royal Soc. Chem., ed N.G. Stanley-Wood and R. Lines, *197*

8   Rayleigh, Lord (1899), *Phil. Mag.*, **48**, 321, *197*

9   Pockels, A. (1891), *Nature,* **43**, 437, *197*
10   Adam N.K. (1941), *The Physics and Chemistry of Surfaces,* Oxford Univ. Press, *197*
11   Linnar, E.R. and Williams, A.P. (1950), *J. Phys. Colloid Chem.,* **54**, 605, *197*
12   Vold, M.J. (1952), *J. Colloid Sci.,* **7**, 196, *197*
13   Kipling, J.J. and Wright, E.H.M. (1963), *J. Chem. Soc.,* 3382, *197, 202*
14   Kipling, J.J. and Wright, E.H.M. (1963), *J. Chem. Soc.,* 3535, *198*
15   Roe, R.-Y. (1975), *J. Colloid Interf. Sci.,* **50**(1), 64–69, *198*
16   McBain, J.D. and Dunn, R.C. (1948), *J. Colloid Sci.,* **3**, 308, *198*
17   Smith H.A. and Hurley, R.B. (1949), *J. Phys. Colloid Chem.,* **53**, 1409, *198, 202*
18   Ward, A.F.H. (1946), *Trans. Faraday Soc.,* **42**, 399, *198*
19   Allen, T. and Patel, R.M. (1971), *J. Colloid Interf. Sci.,* **35**(4), 647–655, *198, 201, 217*
20   Allen, T. and Patel, R.M. (1971), *Particle Size Analysis,* Soc. Anal. Chem, London, *198, 201, 217*
21   Allen, T. and Patel, R.M. (1970), *J. Applied Chem,* **20**, 165–171, *198*
22   Harkins, W.D. and Gans, D.M. (1931), *J. Am. Chem.* Soc., **53**, 2804, *198, 203*
23   Harkins, W.D. and Gans, D.M. (1932), *J. Am. Phys. Chem.,* **36**, 86, *198*
24   Russel, A.S. and Cochran, C.N. (1950), *Ind. Engng. Chem.,***42**, 1332, *198*
25   Freundlich, H. (1907), *Z. Phys. Chem.,* **57**, 385, *199*
26   Linnar, E.R. and Gortner, R.A. (1935), *J. Phys. Chem.,* **39**, 35–67, *199*
27   Traube, I. (1891), *Ann. Phys., Liepzig,* **265**, 27, *199*
28   Holmes, H.N. and McKelvey, J.B. (1928), *J. Phys. Chem.,* **32**, 1522, *199*
29   Langmuir, I. (1917), *Trans. Faraday Soc.,* **42**, 399, *199*
30   Adamson, A.W. (1963), *Physical Chemistry of Surfaces,* Interscience, N.Y. *199, 206, 211, 212*
31   Harkins, W.D. and Dahlstrom, R. (1930), *Ind. Engng. Chem.,* **22**, 897, *199, 200*
32   Boer, J.H. de, (1953), *The Dynamical Characteristics of Adsorption,* Princeton Univ. Press, *200*
33   Berthier, P., Kerlan, L. and Courty, C. (1858), *C. R. Acad. Sci.,* Paris, **246**, 1851, *200*
34   Greenhill, E.B. (1949), *Trans. Farad. Soc.,* **45**, 625, *200*
35   Russell, A.S. and Cochran, C.N. (1950), *Ind. Eng. Chem.,* **42**, 1332, *200*
36   Krasnovskii, A.A. and Gurevich, T.N. (1949), *Chem. Abstr.,* **43**, 728, *200*

37   Innes, W.D. and Rowley, H.H. (1947), *J. Phys. Chem.*, **51**, 1172, *200*

38   Hirst, W. and Lancaster, J.K. (1951), *Trans. Faraday Soc.*, **47**, 315, *200*

39   Weissberger, A. *et al.* (1955), *Organic Solvents*, Interscience, N.Y., *200*

40   Langmuir, I. (1917), *J. Am. Chem. Soc.*, **39**, 1848, *202*

41   Hutchinson, E. (1947), *Trans. Faraday Soc.*, **43**, 439, *202*

42   Gregg, S.J. (1947), *Symp. Particle Size Analysis*, Trans. Inst. Chem. Eng., London, **25**, 40–46, *202*

43   Greenhill, E.B. (1949), *Trans Faraday Soc.*, **45**, 625, *202*

44   Crisp, D.J. (1956), *J. Colloid Sci.*, **11**, 356, *202*

45   Smith, H.A. and Fuzek, J.F. (1946), *J. Am. Chem. Soc.*, **68**, 229, *202*

46   Maron, S.H., Ulevith, I.N.and Elder, M.E. (1949), *Anal.Chem.*, **21**, 691, *202*

47   Hanson, R.S. and Clampitt, B.H. (1954), *J. Phys. Chem.*, **58**, 908, *202, 203*

48   Bartell, F.E. and Sloan, C.K. (1929), *J. Am. Chem.* Soc.,**51**, 1637, *203*

49   Ewing, W.W. and Rhoda, R.N. (1951), *Anal.Chem.*, **22**, 1453, *203*

50   Candler, C. (1951), *Modern Interferometers*, Hilger, London, *203*

51   Groszek, A.J. (1968), *SCI Monograph* No. 28, 174, *203*

52   Orr, C. and Dallavalle, J.M. (1959), *Fine Particle Measurement*, MacMillan, N.Y., *203*

53   Gregg, S.J. (1947), *Symp. Particle Size Analysis*, Trans. Inst. Chem. Eng, London, **25**, 40–46, *204*

54   Hirst, W. and Lancaster, J.K. (1951), *Trans. Faraday Soc.*, **47**, 315, *204*

55   Jenkel, E. and Rumbach, B, (1951), *Z. Electrochem.*, **55**, 612, *204*

56   Habden, J.F. and Jellinek, H.H.G. (1953), *J. Polym. Sci.*, **11**, 365, *204*

57   Flory, P.J. (1953), *Principles of Polymer Chemistry*, Cornhill Univ. Press, Ithaca, p. 579, *204*

58   Morawetz, H. (1965), *Macromolecules in Solution*, Interscience, N.Y., *204*

59   Frisch, H.C. and Simha, R. (1954), *J. Phys. Chem.*, **58**, 507, *204*

60   Jellinek, H.H.G. and Northey, H.L. (1985), *J. Polym. Sci.*, **14**, 583, *204*

61   Eltekov, A.Yu. (1970), *Surface Area Determination*, ed D.H. Everett and R.H. Ottewill, Butterworths pp 295–298, *204*

62   Kolthoff, I.M. and MacNevin, W.N. (1937), *J. Am. Chem. Soc.*, **59**, 1639, *204*

63   Japling, D.W. (1952), *J. Appl. Chem.*, **2**, 642, *204*

64   Giles, G.H., Silva,, A.P. de. and Trivedi, A.S. (1970), *Surface Area Determination* ed. D.H. Everett and R.H. Ottewill, Butterworths, pp 295–298, *204, 205*

65   Giles, C.H. *et al.* (1959), *J. Chem. Soc.*, 535–544, *204*

66   Giles, C.H., Greczek, J.J. and Nakhura, S.N. (1961), *J. Chem. Soc.*, 93–95, *204*

67   Giles, C.H., Easton, I.A.S. and McKay, R.B. (1964), *J. Chem. Soc.*, 4495–4503, *204*

68   Allington, M.H. *et al.* (1958), *J. Chem. Soc.*, **8**, 108–116, *204*

69   Giles, C.H. *et al.* (1958), *J. Chem. Soc.*, **8**, 416–424, *204*

70   Wegman, J. (1962), *Am. Dyest Report*, **51**, 276, *204, 205*

71   Padhye, M.R. and Karnik, R.R. (1971), *Indian J. Technol.*, **9**, 320–322, *204*

72   Giles, C.H. and Trivedi, A.S. (1969), *Chem. Ind.*, 1426–1427, *205, 206*

73   Giles, C.H. *et al.* (1978), *Proc. Conf. Struct. Porous Solids*, Neuchatel, Switzerland, Swiss Chem. Soc., *205, 206*

74   Padday, J.F. (1970), *Surface Area Determination*, ed D.H. Everett and R.H. Ottewill, pp 331–337, Butterworths, *205, 206, 207*

75   Mesderfer, J.W. *et al.* (1952), *TAPPI*, **35**, 374, *205*

76   Lyklema, J. and Hull, H.J. van der (1970), *Surface Area Determination*, ed. D.H. Everett and R.H. Ottewill, pp. 341–354, Butterworths, *206*

77   Clark, J.d'A. (1942), *Paper Trade J.*, **115**, 32, *206*

78   Giles, C.H. and Tolia, A.H. (1964), *J. Appl. Chem.*, **14**, 186–194, *206*

79   Giles, C.H., D'Silva, A.P. de. and Trivedi, A.S., (1970), *J. Appl. Chem.*, **20**, 37–41, *207*

80   Giles, C.H. *et al.* (1971), *J. Appl. Chem. Biotechnica*, **21**, 59, *207*

81   Giles, G.H., Silva, A.P. de.. and Trivedi, A.S. (1970), *Surface Area Determination*, ed D.H. Everett and R.H. Ottewill, Butterworths, pp. 295–298, *207*

82   Spenadel L. and Boudart, M. (1960), *J. Phys. Chem.*, **64**, 204, *207*

83   Adams, C.R., Benesi, H.A., Curtis, R.M. and Meisenheimer, R.G. (1962), *J. Catalysis*, **1**, 336, *207*

84   Hassan, S.A., Khalil, F.H. and el–Gamal, F.G. (1976), *J. Catalysis*, **44**, 5, *207*

85   Mikhail, R.Sh. and Robens, E. (1983), *Microstructure and Thermal Analysis of Solid Surfaces*, p. 112, Heyden, *208*

86   Adnadevic, B.K. and Vucelic, D.R. (1978), *Glas. Hem. Drus.*, Belgrade, **43**(7), 385–392, *208*

87   Larionov, O.G. (1976), *V. sh Adsorbtsiya i Poristot.*, 122–126, *208*

88   Veselov, V,V, and Galenko, N.P. (1974), *Zh. Fiz. Khim.*, **48**(9), 2276–279, *208*

89   Koganovskii, A.M. and Leuchenko, T.M. (1976), *Dopou Akad., Ukr. SSR*, Ser. B, 4, 326–328, *208*

90   Gata, G. (1975), *Tek. Econ. Inst. Geol. Rumania*, Ser. **1**, 13, 13–19, *208*

91   Mazden, F.T. (1977), *Thermokin. Akta.*, **21**(1), 89–93, *208*

92   Kulshreshtha, A.K., Chudasama, V.P. and Dweltz, N.E. (1976), *J. Appl. Polym. Sci.*, **20**(9), 2329–2338, *208*

93   Richter, V., Merz, A. and Morgenthal, J. (1975), *Konf. Met. Proszkow, Pol., Mat.*, Konf Inst. Metal Niezelaz, Glivice, Poland, pp 123–134, *208*

94   Ruzek, J. and Zbuzek, B. (1975), *Silikaty*, **19**(1), 49–66, *208*

95   Kloshko, B.N. *et al.* (1974), *Nauch. Tekhu, sb*, **6**, 22–24, *208*

96   Chetty, K.V. and Naidu, P.R. (1972), *Proc. Chem. Symp.*, **1**, 79–83, Dept. Atom. Energy, Bombay, *208*

97   Corrin, M.L. *et al.* (1949), *J. Colloid Sci.*, **4**, 485–495, *208*

98   Lennard–Jones, J.E. and Devonshire, A.F. (1937), *Proc. Royal Soc.*, 163A, 53, *209*

99   Brillouin, L. (1938), *J. Phys.*, **9**(7), 462, *209*

100   Michand, F. (1939), *J. Chim. Phys.*, **36**, 23, *209*

101   Harkins, W.D. (1919), *Proc. Natl. Akad. Sci.*, **5**, 562, *209*

102   Bangham,, D.H. and Razouk, R.I. (1937), *Trans. Faraday Soc.*, **33**, 1459, *210*

103   Bangham,, D.H. and Razouk, R.I. (1938), *J. Proc. Royal Soc.*, **166**, 572, *210*

104   Razouk, R.I. (1941), *J. Phys. Chem.*, **45**, 179, *210*

105   Chessick, J.J., Young, G.J. and Zetttlemoyer, A.C. (1954), *Trans. Faraday Soc.*, **50**, 587, *211*

106   Taylor, J.A.G. (1965), *Chem. Ind.*, 2003, *211*

107   Harkins, W.D. and Jura, G. (1944), *J. Am. Chem. Soc.*, **66**, 1362, *212*

108   Clint, J.H. *et al.* (1970), *Proc. Int. Symp. Surface Area Determination*, Bristol (1969), Butterworths, *212*

109   Partyka, S. *et al*, (1975), *4th Int. Conf. Thermodynamic Vhem.*, (CR), **7**, 46–55, *212*

110   Partyka, S., Rouquerol, F. and Rouquerol, J. (1979), *J. Colloid Interf. Sci.*, **68**(1), 21–31, *212*

111   Chessick, J.J. and Zettlemoyer, A.C. (1959), *Adv. Catalysis*, **11**, 263, *213*

112   Zettlemoyer, A.C. and Chessick, J.J. (1964), *Adv. Chem. Ser.*, **43**, 88, *213*

113   Zettlemoyer, A.C., Young, G.J., Chessick, J.J. and Healey, F.H. (1953), *J. Phys. Chem.*, **57**, 649, *213*

114   Allen, T. and Burevski, D. (1977), *Powder Technol.*, **17**(3), 265–272, *216*

115   Sips, R. (1950), *J. Chem. Phys.*, **18**, 1024, *216*

116   Allen, T. and Burevski, D. (1977), *Powder Technol.*, **18**(2), 139–148, *217*
117   Allen, T. and Burevski, D. (1978), *Powder Technol.*, **21**(1), 91–96, *217*
118   Patel, R.M. (1971), *Physical Adsorption at Solid–Liquid Interfaces*, PhD Thesis, Univ. Bradford, U.K., *217*
119   Burevski, D. (1975), PhD Thesis,, Univ. Bradford U.K., *217*
120   Mikhail, R.Sh. and Robens, E. (1983), *Microstructure and Thermal Analysis of Solids*, John Wiley, *217*
121   Kuelen, J. van. (1973), *Material and Construction*, **6**(33), 181, *217*
122   Tuul, J. and Innes, W.B. (1962), *Anal. Chem.*, **34**(7), 818–820, *218*
123   Jäkel, K. (1972), *Beckmann Report*, **2**, S33–35, Beckmann Instruments, *218*

Appendix

Names and addresses of manufacturers and suppliers

AB Atomenergie, Studvik, Sweden.

Adams, L. Ltd, Minerva Road, London NW10, UK

Addy Products Ltd, Solent Industrial Estate, Botley, Hampshire SO3 2FQ, UK

Advanced Polymer Systems, 3636 Haven Road, Redwood City, CA 94063, USA (415) 366 2626

Aerograph Co., Lower Sydenham, London SW26, UK

Aerometrics Inc., 550 Del Rey Avenue, Sunnyvale, CA 94086, USA, (408) 738 6688

Agar, Alan W., 127 Rye Street, Bishop's Stortford, Hertfordshire, UK

Air Supply International, Gateway House, 302–8 High Street, Slough, Berks, UK

Air Techniques Inc., 1717 Whitehead Road, Baltimore, Maryland 21207, USA

Airflow Development, 31 Lancaster Road, High Wycombe, Bucks, UK

Allied Scientific Company Ltd, 2220 Midland Avenue, Scarborough, Canada

Alpine 89, Augsburg 2, Postfach 629, Germany

Alpine American Corporation, 5 Michigan Drive, Natick, MA 01760, USA, (617) 655 1123

Ameresco Inc., 101 Park Street, Montclair, NJ 07042, USA

American Innovision, 9581 Ridgehaven Court, San Diego, CA 92123, USA, (619) 560 9355

American Instrument Co., 8030 Georgia Avenue, Silver Springs, MD 20910, USA

Amherst Process Instruments Inc., Mountain Farms Technology Park, Hadley, MA 1035-9547, USA, (413) 586 2744

Amtec, Alpes Maritimes Technologie, 1er C.A., Avenue du Docteur, Julien-Lefebvre, 06270 Villeneuve-Loubert, France.(93 73 40 20)

Analytical Measuring Systems, London Road, Pampisford, Cambridge, Cambs. CB2 4EF, England (01223) 836001

Analytical Products Inc., 511 Taylor Way, Belmont CA 94002, USA, (415) 592-1400

Andersen Instruments Inc., Graseby-Andersen, 4801 Fulton Ind. Blvd, Atlanta, GA 30336-2003, USA, (404)691 1910

Anderson 2000 Inc., PO Box 20769, Atlanta, Georgia 30320, USA

Anotec Separations, (Membrane Filters), 226 East 54th Street, New York, NY 10022, USA, (212) 751 0191

Anton Paar USA Inc., 340 Constance Dr., Warminster, PA 18974, USA, (215) 443 7986

Applied Research Laboratories, Wingate Road, Luton, Beds, UK.

Armco Autometrics, 7077 Winchester Circle, Boulder, Colorado, 80301, USA, (303) 530 1600

Artec Systems Corp. 170 Finn Court Farmingdale, NY 11735, USA, (516) 293 4420, [France: Nachet, 106 Rue Chaptal, 92304 Levallois Peret— Cedex, 757 31 05]

Artek Systems Corporation, 170 Finn Ct., Farmingdale, N.Y. 11835, USA, (713) 631 7800

Atcor Instrumentation Division, 2350 Charleston Road, Mountain View, CA 94043, USA (415) 968 6080

Ateliers Cloup, 46 Boulevard Polangis, 94500, Champigny-sur-Marne, France

ATM Corporation, 645 S. 94th Pl., Milwaukee, WI 53214, USA, (414) 453 1100

ATM Corporation, Sonic Sifter Division, 645 S. 94th Pl., Milwaukee, W., 53214–1206, USA, (414) 453 1100

ATV Gaulin Inc., 44 Garden Street, Everett, MA 02149, USA, (617) 387 9300

AWK Analyzers, Contamination Control Systems (CCC), Gaigistrasse 3, D-800 Munich 2, Germany, (089) 188006/07

Babcock and Wilcox Ltd, Cleveland House, St James's Square, London SW14 4LN, UK

Bailey Meter Company, Wickliffe, Ohio 44092, USA

Bailey Meters and Controls, 218 Purley Way, Croydon, Surrey,UK

Bausch & Lomb Inc., 820 Linden Avenue, 30320 Rochester, NY 14625, USA, (716) 385 1000

BCR, (Community Bureau of Reference), Directorate General X11, Commission of the European Communities, 200 rue de la Loi, B-1049, Brussels, Belgium

Beckmann Instruments Inc., Fullerton, CA 92634, USA

Beckmann Instruments Ltd, Glenrothes, Fife, Scotland

Bekaert NV, SA, Bekaerstraat 2, B8550 Zwevegem, Belgium. 32-56 766292

Bel Japan, Inc., 1-5 Ebie 6-chome, Fukushima-ku, Osaka 553, Japan, 06-454-0211

Belstock Controls, 10 Moss Hall Crescent, Finchley, London N12 8NY,UK, (0181) 446 8210

Bendix Corporation, Env. Sci. Div., Dept 81, Taylor Ave., Baltimore, MD, 21204, USA

Bendix Vacuum Ltd, Sci. Instr. & Equ. Div., Easthead Ave., Wokingham, Berks, RG11 2PW, UK

Berkley Instruments Inc., 2700 Dupont Drive, Irwine, CA 92715, USA

Beta Scientific Corp PO Box 24, Albertson, New York 11507, USA, (516) 621 7971

Biophysics, Baldwin Place Road, Mahopac, NY 10541, USA

Biorad Laboratories, 2200 Wright Ave., Richmond, CA 94804, USA, (415) 234 4130

Boeckeler Instruments Inc., 3280 East Hemisphere Loop, #114, Tucson, AZ 85706-5024, USA, (602) 573 7100

Brezina, J., Hauptstrasse 68, D6901 Waldhilsbach, Germany

Brinkmann Instrument. Company, 1 Cantiague Road, PO Box 1019, Westbury, NY 1159-0207, USA, (516) 334 7506

Bristol Industrial Research Associates Ltd, (BIRAL), PO Box 2, Portishead, Bristol, S20 9JB, UK, 275 847303

British Rema, PO Box 31, Imperial Steel Works, Sheffield S9 1RA, UK

Brookhaven Instrument Corporation, 750 Blue Point Road, Holtsville, NY 11742, USA, (516) 758 3200

Buck Scientific Inc., 58 Fort Point Street, E. Norwalk, CT 06855, USA, (203) 853 9444

Buckbee Mears Co., 245 East 6th Street, St Paul,, MN 55101, USA, (612) 228 6400

Buehler, Optical Instruments Division, 41 Waukegan Road, Lake Bluff, IL, USA, 60044 (312) 295 6500

Cahn Instruments Inc., 16207 South Carmenita Road, Cerritos, CA 90723, USA, (213) 926 3378

Cahn, 27 Essex Road, Dartford, Kent, UK

California Measurements Inc., 150 E.Montecito Avenue, Sierra Madre, CA 91024, USA, (818) 355 3361

Cambridge Instruments Inc.,Viking Way, Bar Hill, Cambridge, U.K., (0954) 82781

Cambridge Instruments Ltd, PO Box 123, Buffalo New York 14240, USA, (716) 891 3000

Cambridge Instruments Ltd, Viking Way, Bar Hill, Cambridge, CB3 8EL, U.K., (0954) 82020

Cambridge Instruments Sarl,Centre d'Affaires Paris Nord, 93153 Le Blanc Mesnil, France. (1 4867 01 34)

Canty, J.M. Assoc. Inc., 590 Young Street, Tonawanda, NY 14150,USA, (716) 693 3953

Cargill. R. P., Laboratories Inc., Cedar Grove, NJ, USA

Carl Zeiss Jena Ltd, England House, 93-7 New Cavendish Street, London W1,UK

Carl Zeiss, 444 Fifth Avenue, New York. NY 10018, USA

Carl Zeiss, 7082 Oberkochen, Germany

Carlo Erba, via Carlo Imbonati 24, 20159 Milan, Italy [USA Rep: Haake Buchler, UK Rep: Casella, C. F. & Co.,Regent House, Britannia Walk, London N1 7ND]

Celsco Industries Ltd, Environ. & Ind. Products, Costa Mesa CA, USA

Central Technical Institute, CTI-TNO, PO Box 541, Apeldoorn, The Netherlands

Chandler Engineering Co, 707 E38 Street, Tulsa, OK.74145, USA, (918) 627 1740

Charles Austin Pumps, Petersham Works, 100 Royston Road, Byfleet, Surrey, UK

Chemische Laboratorium für Tonindustrie, Goslar, Harz, Germany

Christison, A. (Scientific Equipment) Ltd, Albany Road, East Gateshead Industrial Estate, Gateshead 8, UK

Cilas Alcatel, Granulometry Dept., Route de Nozay, BP 27, F-91460 Marcoussis, France, (33) 16454-48 00

Clay Adams, Div. of Becton Dickinson, Parsippany, NJ, USA

Climet Instruments Co., 1320 W. Colton Ave., PO Box 1760, Redlands, CA 92373,USA, (714) 793 2788

Coleman Instruments Inc., 42 Madison Street, Maywood, IL, USA

Compagnie Industrielle des Lasers, Route de Nozay, 91 Marcoussis, France

Compix Inc. Imaging Systems, 230 Executive Drive, Suite 102-E, Mars, PA 16046, USA, (412) 772 5277

Contamination Control Systems, Gaigistasse 3, München 2, Germany, (089) 18 80 06/07

Contest Instruments Ltd, Downmill Road, Bracknell, Berkshire RG12 1QE, UK

Core Laboratories Inc., Div. of Litton, 7501 Stemmons, Dallas, TX 75247, USA, (214) 631 8270

Coulter Electronics Ltd, Northwell Drive, Luton, Beds., LU3 3RH, UK, (0582 491414)

Coulter Inc., Box 2145, 590 W.20th Street, Hialieah, FL 33010, USA, (305) 885 0131

CSC Scientific Company, Fairbanks, VA, USA, (703) 876 4030

Curtin Matheson Scientific Inc., 9999 Veterans Memorial Dr., Houston, TX 77038, USA, (713) 878 2349

Danfoss Vision, Jegstrupvej 3, DK-8361 Hasselager, Denmark, 45 89 48 93 88

Danfoss Interservices GmbH, Carl-Legien Strasse 8, 63073 Offenbach/Main, Germany, (069) 89 02 177

Dantec Electronics Inc., 177 Corporate Drive, Mahwah, NJ 07430,USA, (201) 512 0037

Dantec Elektronic, Medicinsk og Videnstkabeligt Maleudstyr A/S, Tonsbakken 16–18, DK 2740 Skovlunde, Denmark

Data Translation, 100 Locke Drive, Marlboro, MA 01752-1192, USA, (508) 481 3700

Datametrics Div., CGS Scient. Corp., 127 Coolidge Hill Rd., Watertown, MA 02172, USA

Day Sales Company, 810 Third Ave. NE, Minneapolis 13, MN, USA

Degenhardt & Co. Ltd, 6 Cavendish Square, London WI, UK

Del Electronics Corp., 616-T Adams Street, Steubenville, OH 43952,USA

Delaran Manufacturing Co., West Des Moines, USA

Delcita Ltd, Ver House, London Road, Markyate, Herts, AL3 8JT, UK, 01582 841665

Delviljem (London) Ltd, Delviljem House, Shakespeare Road, Finchley, Middlesex, UK

Denver Process Equipmennt Ltd, Stocks House, 9 North Street, Leatherhead, Surrey KT22 7AX, UK, 01372 379313

Denver Instrum. Co., Ainsworth Div., 2050 South Pecos Street, Denver, CO, 80223, USA

Diagnetics Inc, 5410 South 94th East Avenue, Tulsa, 74145-8118, USA, (918) 664 7722

Dietert, H. & Company, 9330 Roselawn Ave., Detroit, MI, USA

Dixon, A.W. & Co., 30 Anerly Station Road, London SE20, UK

Donaldson Company Inc., 1400 West 94th Street, Minneapolis, MN 55431, USA

Dow Chemicals, Midland, MI, USA

Draeger Normalair Ltd, Kitty Brewster, Blythe, Northumberland, UK

Drägerwerk Lubeck, D-24 Lubeck 1, PO Box 1339, Moislinger Allee 53–55, Germany

Duke Scientific Corp., 1135D San Antonio Road, PO Box 50005, Palo Alto, CA 94303, USA, (415) 424 1100

DuPont de Nemours, E.I., (Dillon F. Schofield), DuPont Imaging R&D, Glasgow Community MS603, PO Box 6110, Newark, DE 19714-6110, USA

DuPont Company (E.I.Dupont de Nemours) Materials Characterisation Systems, Concord Plaza, Quillen Building, Wilmington, DE 19898, USA. (302) 772 5488.

Dynac Corporation, Thompsons Point Portland, ME, USA

Dynac Div. of Dieldstone Corp., PO Box 44209, Cincinnati, OH, USA

Ealing Beck Ltd, Greycaine Rd., Watford WD2 4PW,UK

Eberline Instruments Corp., PO Box 2108, Sante Fe, NM 87501, USA

EDAX International Inc.,103 Schelter Rd, PO Box 135, Prairie View, IL 60069, USA, (312) 634 0600

Edison, Thomas A. Industries, Instruments Div., West Grange NJ, USA

Electronics Design A/S, Chr, Holms Parkvej 26, 2930 Klampenborg, Denmark

Endecottes Ltd, 9 Lombard Road, London, SW19 3BR, UK, (081) 542 8121, [US Distributer, CSC Scientific Company]

Engelhardt Industries Ltd, Newark, NJ, USA

Environeering Inc., 9933 North Lawler, Skokie, IL 60076, USA

Environmental Control International (ECI) Inc., 409 Washington Ave. PO Box 10126, Baltimore, MD, USA

Environmental Monitoring Systems Ltd, Kingswick House, Kingswick Drive, Sunninghill, Berkshire SL5 7BH, UK, (01990) 23491

Environmental Research Corp. (ERC), 3725N. Dunlap St.,St Paul, MN 55112, USA

Erba Instruments Inc.,4 Doulton Place, Peabody, MA 01960, USA, (617) 535 5986

Erba Science (UK) Ltd, 14 Bath Street, Swindon, SN1 4BA, UK

Erdco Engineering Corp., 136 Official Road, Addison, IL 60101, USA, (708)328 0550

Erwin Sick Optik-Elektronik, D-7808 Waldkirch, An der Allee 7-9 Posfach 310, Germany

Evans Electroselenium Ltd, Halstead, Essex, UK

Fairy Ind. Ceram., Filleybrooks, Stone, Staffs., ST15 OPU, UK, 01785 813241

Faley International Corporation, PO Box 669, El Toro, CA 92630-0669, USA, (714) 837 1149

Fawcet Christie Hydraulics, Sandycroft Ind. Estate, Chester Road, Deeside, Clwyd, UK, 01244 535515

Fffractionation Inc., 1270 W. 2320 South, Suite. D., Salt Lake City, UT 84119, USA, (801) 975 7550

Ficklen, Joseph B., 1848 East Mountain Street, Pasadena 7, CA 91104, USA

Filtra GmBh, Filtrastrasse 5-7, D-4730 Ahlen 5, Germany, 49 2528 300

Fisher Scientific Co., 711 Forbes Ave., Pittsburg 19, PA 15219, USA, (412) 787 6322

Fisons Instruments S.p.A., Strada Rivoltana–20090, Rodano, Milan, Italy, 2–9505 9272

Fleming Instruments Ltd, Caxton Way, Stevenage, Herts, UK

Flowvision, Kelvin Microwave Corp., Charlotte, NC, USA, (704) 357 9849

Foxboro Company,Bristol Park, Foxboro, MA 02035, USA, (617) 543 8750

Fortress Dynamics, 25 Ladysmith Road. Gloucester, GL1 5EP, UK, (01452 305057)

Foster Instruments, Sydney Road, Muswell Hill, London N10, UK

Franklin Electronics Inc., Bridgeport, PA, USA

Freeman Labs. Inc., 9290 Evenhouse Avenue, Rosemount, IL 60018, USA

Fritsch, Albert & Co., Industriestrasse 8, D6580 Idar-Oberstein 1, Germany, 06784/70 0

Galai Instruments, Inc., 577 Main Street, Islip, New York 11751, USA, (516) 581 8500

Galai Production, PO Box 221, Industrial Zone, Migdal, Haemek 10500, Israel, 972–654 3369

Gallenkamp Ltd, Portrack Lane, Stockton on Tees, Co. Durham, UK

Gardner Association Inc., 3643 Carman Road, Schenectady, NY 12303, USA

Gardner Laboratory, Bethesda, MD, USA

GCA Corporation, 213 Burlington Road, Bedford, MA, 01730, USA, (617) 275 5444

Gelman Hawksley, 12 Peter Road, Lancing, Sussex, UK

Gelman Instruments Co., 600 South Wagner Road, Ann Arbor, MI 48106, USA

Gelman Sciences Ltd, Brackmills Business Park, Caswell Road, Northampton NNA OEZ, UK, 01604 765 141

Gelman Sciences, 600 S Wagner Road, Ann Arbor, MI 48106, USA, 313 665 0651.

General Electric Co., Schenectady, NY, USA

General Electric Ordinance Systems, 100 Plastic Ave., Pittsfield, MA 01201, USA

General Motors Corp., Flint, MI, USA

General Sciences Corp., Bridgeport, CT 06604, USA

Gerber Scientific Inc., 1643 Bentana Way, Reston, PO Box 2411, VA. 22090, USA, (703) 437 3272

Gilson Co. Inc., PO Box 677, Worthington, OH 43085-0677, USA, (614) 548 7298

Giuliani , via Borgomanero 49, Turin, Italy

Glass Developments Ltd, Sudbourne Road, Brixton Hill, London, SW1,UK

Glen Creston, 16 Dalston Gardens, Stanmore, Middlesex HA7 1DA ,UK, 0181 206 0123

Global Lab., 100 Locke Drive, Marlboro, MA 01752-1192, USA

Goring Kerr Ltd, Hanover Way, Windsor, Berkshire, UK

Gould Inc., Design & Test Systems Div., 4650 Old Ironsides Drive, Santa Clara, CA 95054-1279, USA, (408) 988 6800

Graticules Ltd, Morley Road, Tonbridge, Kent TN9 1RN (01732) 359061,UK

Greenfield Instruments, Div. of Beta Nozzle, Greenfield, MA, USA, (413) 772-0846

Greenings, Britannia Works, Printing House Lane, Hayes, Middlesex, UK

Griffin & George Ltd, Wembley, Middlesex, UK

Gustafson Inc., 1400 Preston Rd., Plano, Texas, USA, (214) 985-8877

Haake Buchler Instruments Inc., 244 Saddle River Road, Saddle Brook, NJ 07662-6001, USA, (201) 843 2320

Hamamatsu Photonics France, 49/51 Rue de la Vanne, 92120 Montrouge, France 46 55 47 58

Hamamatsu Systems Inc., 332 Second Avenue, Waltham, MA 02154, USA, (617) 890 3440

Hammatsu Corp.,120 Wood Ave., Middlesex, NJ 08846, USA

Harrison Cooper Associates, Salt Lake City, Utah, USA

Hawksley & Sons Ltd, 12 Peter Road, Lancing, Surrey, UK

Hiac/Royco Div., Pacific Scientific, 11801 Tech Road, Silver Springs MD 20904 (301) 680 7000

High Yield Technology (HYT), 800 Maude Ave., Mountain View, CA 94043, USA, (415) 960 3102

Hird-Brown Ltd, Lever Street, Bolton, Lancashire, BL3 6BJ, UK

Hitech Instruments Inc., PO Box 886, 4799 West Chester Pike, Edgemont, PA 19028, USA, (215) 353-3505.

Horiba France, 13 Chemin de Levant, 01210–Ferney-Voltaire, France, 33 50 40 85 38

Horiba Instrum. Corp.,17671 Armstrong Ave, Irvine, CA 92714, USA (714) 250 4811

Horiba Instrum. Ltd,5 Harrowden Road, Brackmills, Northhampton, NN4 0EB, UK, (01604 65171)

Horiba Instruments Inc., 17671 Armstrong Ave., Irvine, CA 92714, USA, (714) 250 4811

Hosokawa Micron International Inc., 10 Chatham Road, Summit, NJ 07901, USA, (908) 598 6360

Howden Wade Ltd, Crowhurst Road, Brighton, Sussex BNI 8AJ, UK, 01273 506311

Howe, V.A. & Co. Ltd, 88 Peterborough Rd.,London SW6, UK

Hydrosupport, Porsgrum, Pb 2582, Hydro 3901, Porsgrunn, Sweden, 47 3556 30 05

ICS, 91 rue du General de Gaulle, BP1, 27109 Le Vaudreuil, Cedex, France, 33 32 09 36 26

Image Analysing Computers Ltd, Melbourne Rd., Royston, Herts. SG6 6ET, UK

Imanco, 40 Robert Pitt Dr., Monsey, NY 10952, USA

Imperial Chemical Industries Ltd, Nobel Div., Stevenston, Ayrshire, Scotland, UK

Independent Equipment Corp., PO Box 460, Route 202N, Three Bridges, NJ 08887, USA, (201) 782 5989
Industri Textil Job AB, Box 144, S-51122, Kinna, Sweden, 46 320 13015
Infrasizers Ltd, Toronto, ON, Canada
Insitec Inc.,2110 Omega Rd., Suite D, San Ramon, CA 94583, USA, (510) 837 1330
International Trading Co. Inc., 406 Washington Ave.,P O Box 5519, Baltimore, MD 21204, USA
International Trading Co., Orchard House, Victoria Square, Droitwich, Worcestershire WR9 8QT, UK
InterSystems Industrial Products, 17330 Preston Road, Suite 105D, LB 342, Dallas, TX 75252, USA

Japan Electron Optics Ltd, Jeolco House, Grove Park, Edgeware Rd., Colindale, London NW9, UK
Japan Electron Optics, 11 Dearborn Rd., Peabody, MA 01960, USA, (508) 535 5900
Jenoptic Technologie GmbH, Unternehmensbereich Optische Systemtechnik, D–07739 Jena, Germany, (0 36 41) 65 33 27
Joy Manufacturing Co., Western Precipitation Div., 100 West 9th Street, Los Angeles, CA 90015, USA
Joyce-Loebl, Marquisway, Team Valley, Gateshead NE11 0QW, UK, (0191) 482 2111

Kane May Ltd, Northey International Division, Nortec House, Chaul End Lane, Luton, Beds. LU4 8E2, UK, 01582 584343
Kek Ltd,Hulley Road, Hurdsfield Ind. Est., Macclesfield, Cheshire SK10 2ND, UK
Kelvin Microwave Corp., (Flowvision), Charlotte, NC., USA, (704) 357 9849
Kevex Corporation, 1101 Chess Drive, Foster City, CA, USA, 94404 (415) 573 5866
Kevex Instruments, 355 Shoreway Road, PO Box 3008, San Carlos, CA 94070-1308, USA (415) 591 3600.
Kevex UK, 37 Alma Street, Luton, Beds., LU1 2PL, UK, 01582 400596
KHD Industrie-Anlagen GmbH, D-5000, Köln, Germany
Kontron D-8 München 50, Lerchenstrasse 8-10, München, Germany
Kontron Elektronik (Zeiss) GmbH, Image Analysis Division, Breslauer Str. 2, 8057 Etching/München
Kowa Optimed Inc., 20001 South Vermont Avenue, Torrence, CA 90502, USA, (213) 327 1913
Kratel SA, CH-1222 Geneve-Vesenaz, 64 Ch de St Maurice, Switzerland, 022 52 33 74

Labcon Ltd, 24 Northfield Way, Aycliffe Industrial Estate, Newton Aycliffe, Co. Durham DL5 6EJ, UK, (01325) 313379
La Pine Scientific Co., Chicago 29, IL, USA
Lars A.B., Ljungberg & Co., Stockholm, Sweden

Lasentec, Laser Sensing Technology Inc., 15224 NE 95th Street, Redmond, WA, 98052, USA, (206) 881 7117)

Laser Associates Ltd, Paynes Lane, Warwickshire, UK

Laser Holography Inc., 1130 Channel Drive, Santa Barbara, CA 93130, USA

Laser Lines Ltd, Beaument Close, Banbury, Oxon. OX16 7TQ, UK, (01295) 67755

Lavino, Garrard House, 31-45 Gresham Street, London EC2, UK

Leco Corporation, 3000 Lakeview Avenue, St. Joseph, MI 49085–2396, USA, (616) 983 5531

Leeds & Northrup Instruments, Sunneytown Pike, PO Box 2000, North Wales, PA 19454, USA., (215) 699 2000

Leica Inc., 111 Deer Lake Road, Deerfield, IL 60015, USA

Leico UK Ltd, Davy Avenue, Knowhill, Milton Keynes, MK5 8LB, UK, (01908) 666663

Leitz, Ernst, D-633 Wetzlar GmbH, Postfach 210, Germany

Leitz, Ernst, Rockleigh, NJ 07647, USA

Leitz, Ernst, 30 Mortimer Street, London W1N 8BB, UK

LeMont Scientific, Inc. 2011 Pine Hall Drive, Science Park, State College, PA 16801, USA, (814) 238 8403)

Lindley Flowtech Ltd, 895 Canal Road, Frizinghall, Bradford BD2 1AX, UK, 01274 530066

Litton Systems Inc., Applied Science Div., 2003 E. Hennegin Ave., Minneapolis, MN 55413, USA

M & M Process Equipment Ltd, Fir Tree House, Headstone Drive, Wealdstone, Harrow, Middlesex HA3 5QS, UK

Macro Technology, Kingswick House, Kingswick Drive, Sunninghill, Berks SL5 7BH, UK, 01990 23491

Malvern Instruments Inc., 10 Southville Rd, Southborough, MA 01772, USA., (508) 480 0200

Malvern Instruments Ltd, Spring Lane, Malvern, Worcs.,WR14 1AL, UK, (016845 3531)

Manufacturing Engineering & Equipment Corp., Warrington, PA, USA

Mason & Morton Ltd, 32–40 Headstone Drive, Wealdstone, Harrow, Middlesex, UK

Matec Applied Sciences, 75 South Street, Hopkinton, MA 01748, USA, (508) 435 9039

Matelem, Les Cloviers, rue d'Argenteuil, 95110, Sannois, France

McCrone Research Institute, 2820 S. Michigan Ave., Chicago, IL 60616, USA. (312) 842 7100

Met One, 481 California Avenue, Grants Pass, OR 97526, USA. (503) 479 1248)

Metals Research Ltd, 91 King Street, Cambridge, UK

Metamorphis: Contrado Luchiano 115/C, Zona Industriale, 70123 Modugno, (Bari), Italy, (Tlx 810202)

Meteorology Research Inc., 474 Woodbury Road, Alteydena, CA 91001, USA

Metronics Associates Inc., 3201 Porter Drive, Stamford Industrial Park, PO Box 637, Palo Alta, CA, USA

Mettler (Switzerland) Instruments A.G., CH-8606, Greifensee-Zurich, Switzerland
Micro Measurements Ltd, Shirehill Industrial Estate, Shirehill, Saffron Walden, Essex CB11 3AQ, UK,.(01223 834420) [France, Securite Analytique S.A., 4 Rue Sainte Famille, 78000 Versailles, (953 4609)]
Micromeretics, One Micromeretics Dr., Norcross, GA 30093-1877, USA, (404) 662-3969
Micron Powder Systems, 10 Chatham Road, Summit, NJ 07901, USA, (908) 273 6360
Micropul Corporation, 10 Chatham Road, Summit, NJ 07901, USA, (201) 273 6360
Micropure Systems Inc., 2 Oakwood Place, Scarsdale, NY 10583, USA, (401) 231 9429
Microscal Ltd, 79 Southern Row, London W10 5AL, UK, (0181) 969 3935
Millipore (U.K.) Limited, Millipore House, Abbey Road, London NW10 7SP, UK, (01965) 9611/4
Millipore Corporation, 80 Ashby Road, Bedford, MA 01730, USA, (617) 275 9200
Mines Safety Appliances Co. Ltd, Greenford, Middlesex, UK
Mines Safety Appliances, 201 Braddock Ave., Pittsburg 8, PA, USA
Mintex Div., Cartner Group Ltd, Stirling Rd Trading Estate, Slough, Bucks, UK
MM Industries, (see Vorti-Siv)
Monitek Technologies, Inc., 1495 Zephyr Avenue, Hayward, CA 94544, USA, (415) 471 8300
Monsanto Company, Eng. Sales Dept., 800N Lindberg Boulevard, St Louis, MS 63166, USA
MSA, (Mines Safety Appliances), 201 Braddock Ave., Pittsburgh 8, PA, USA
Mullard Equipment Ltd, Manor Road, Crawley, Sussex, UK
Munhall Company, 5655 High Street, Worthington, OH, 43085, USA, (614) 888 7700

Nachet Vision SA, 125 Boulevard Davout B.P. 128, 75963 Paris Cedex 20, France, 33(1) 43.48.77.10
Nachet Vision [Tegal Scientific Inc., PO Box 5905, Concord, CA 94524, USA, (415) 827 1054
National Bureau of Standards, Washington, USA
National Physical Laboratory, Metrology Div., Teddington, Middlesex, TW11 0LW, UK, (01977 Ext. 3351)
Nautamix, N.V., PO Box 773, Haarlem, Holland
Nebetco Engineering, 1107 Chandler Avenue, Raselle, NJ 07203, USA
Netzch-Gerätebau GmbH, Wittelsbacher Str. 42, D-8672, Selb, Germany
Nethreler & Hinz., GmbH, Hamburg, Germany
NEU Engineering Ltd, 32–34 Baker Street, Weybridge, Surrey, UK
NEU Etablishment, PO Box 28, Lille, France
Ni On Kagaka Kogyo Co. Ltd, 4168 Yamadashimo, Suita, Osaka, Japan
Nicomp Particle Sizing Systems, 6780 Cortona Dr., Santa Barbara, CA 93117, USA, (805) 9681497

Nisshin Engineering Co. Ltd, Tomen America Inc., 1000 Corporate Grove Dr., Buffalo Grove, IL 60089 4507, USA, (708) 520 9520

Nitto Computer Application Ltd, Shibuya 3-28-15, Shibuya-Ku, Tokyo, Japan, 03–498–1651

Normandie–Labo, 76210 Lintot, France

Northey International Systems Ltd, 5 Charles Lane, St John's Wood, High Street London, NW8 7SB, UK

Northgate Traders Ltd, London EC2, UK

Nuclear Enterprises Ltd, Sighthill, Edinburgh 11, Scotland, UK

Nuclear Measurements Corporation, 2460 N. Arlington Ave., Indianapolis, IN, USA

Nuclepore Corporation, 7035 Commerce Circle, Pleasanton, CA 94566-3294, USA, (415) 463-2530

Numek Instruments & Controls Corporation, Appolo, PA, USA

Numinco, 300 Seco Road, Monroeville, PA 15146, USA

Olympus Precision Instruments, 4 Nevada Dr., Lake Success, NY 11042-1179, USA, (516) 488 3880

Oncor Instrument Systems, 9581 Ridgehaven Court, San Diego, CA 92123, USA, (619) 560 9355

Optomax, 9 Ash Street, Hollis, NH 03049, USA, (603) 465 3385

Optomax, 109 Terrace Hall Avenue, Burlington MA.01803, USA, (617) 272 0271

Optronics International Inc., Chelmsford, MA, USA

Otsuka Electronics Co. Ltd, 3-26-3 Shodai-Tajika, Hirakata, Osaka 5873, Japan, 0720-55–8550

OutoKumpu Electronics [Princeton Gamma-Tech, Inc., 1200 State Road, Princeton, NJ 08540, USA, (609) 924 7310]

Outokumpu Mintec Oy, PO Box 84, SF-02201, Espoo, Finland, 358 0 4211

Paar USA Inc., 340 Constance Drive, Warminster, PA 18974, USA, (215) 443 7570

Pacific Scientific Co., Hiac/Royco Div., 11801 Tech. Rd., Silver Springs, MD 20904, USA, (301) 680 7000

Page (Charles) & Co. Ltd, Acorn House, Victoria Rd., London W3 6XU, UK

Palas GmbH, Greschbachstr 36, D-7500 Karlsruhe, Germany, 721 962130

Paris–Labo, 49 rue de France, 94300, Vincennes, France

Parker Hannifin plc, Peel Street, Morley, Leeds LS27 8EL, U.K. 01532 537921

Particle Data Inc., PO Box 265, Elmhurst, IL 60126, USA, (708) 832 5653

Particle Data Ltd, 39 Tirlebank Way, New Town, Tewkesbury, Glos. GL20 5RX, UK

Particle Information Services Inc., PO Box 702, Grant Pass, OR 97526, USA

Particle Measuring Systems Inc., 1855 South 57th Court, Boulder CO 80301, USA, (303) 443 7100

Particle Sizing Systems, 75 Aero Camino, Santa Barbara, CA 93117, USA, (805) 968 0361

Particle Technology Ltd, PO Box 173, Foston, Derbyshire, DE65 5NZ, UK, 01283 520365

Partikel Messtechnik GmbH, (PMT), Carl-Zeiss-Str. 11, Postfach 15 16, D-7250 Leonberg-Gebersheim, Germany, (71 52/5 10 08)

Pascall Ltd, Gatwick Road, Crawley, Sussex, RH10 2RS, UK

Pearson Panke Ltd, 1–3 Halegrove Gardens, London NW7, UK

Pen Kem Inc., 341 Adams St., Bedford Hills, NY 10507, USA, (914) 241 4777

Penwalt Ltd, Doman Road, Camberly, Surrey, UK

Perkin Elmer Ltd, Post Office La., Beaconsfield, Buckinghamshire, HP9 1QA, UK, 014967 9331

Perrier et Cie, rue Marie-Debos, 92120 Montrouge, France

Phoenix Precision Instruments, Gardiner, New York, USA

Photal, Otsuka Electronic Co Ltd, 3-26-3 Shodai-Tajika. Hirakata, Osaka, 573, Japan, (0720 55 8550) [US Distributors Munhall]

Photoelectronics Ltd, Arcail House, Restmor Way, Hockbridge, Wallington, Surrey, UK

PMT Partikel-Messtechnik GmbH, Carl Zeiss Str 11, Postfach 15 16, D-7250 Leonberg-Gebersheim, Germany, 7152 51008

Pola Laboratories Supplies Inc., New York 7, USA

Polaron, 4 Shakespeare Road, Finchley, London N3, UK

Polytec GmbH & Co., D-7517 Waldbronn, Karlsruhe, Germany, (07243 6 99 44) [UK Distributor, Laser Lines]

PPM Inc.,11428 Kingston Pike, Knoxville, TN 37922, USA, (615) 966 8796 [UK Distributor, Macro Technology]

Princeton Gamma-Tech Inc.,1200 State Road, Princeton, NJ 08540, USA, (609) 924 7310

Proassist Company, 1614 East 5600 South, Salt Lake City, UT 84121, USA

Procedyne Corporation, 11 Industrial Dr., New Brunswick, NJ 08901, USA, (908) 249 8347

Process & Instruments Corporation, Brooklyn, New York, USA

Production Sales & Services Ltd, New Malden, Surrey, UK

Prolabo (France), 12 rue Pelee, 75011, Paris X1, France

Prosser Scientific Instruments Ltd, Lady Lane Industrial Estate, Hadleigh, Ipswich, UK

Quantachrome Corp., 1900 Corporate Drive, Boynton Beach, FL 33426, USA, (407) 731 4999

Radiometer-America Inc., 811 Sharon Drive, Westlake, OH 44145, USA, (216) 871 5989

Rao Instruments Co. Ltd, Brooklyn, New York, USA

Rattreyon Learning Systems, Michigan City, IN, USA

Research Appliance Co., Route 8, Gibsonia, PA 15044, USA

Retsch F. Kurt GmbH & Co KG, Reinische Strasse 36, D-5657 Haan 1, POB.1554, Germany (02129) 55 61-0; 9, (UK Representative, Glen Creston)

Reynolds & Branson Scientific Equipment, Dockfield Road, Shipley, Yorkshire, UK.

Rion Co. Ltd, 20–41 Higashimotomachi 3-chome, Kokubunji, Tokyo 185, Japan, (0423-22-113)

Ronald Trist Controls Ltd, 6–8 Bath Rd., Slough, Berks, UK

Rotex Inc.,1230 Knowlten Street, Cincinnati, OH 45223–1845, USA, (513) 541 4888

Rotheroe & Mitchel Ltd, Aintree Rd., Greenford, Middlesex UB6 7LJ, UK

Rupprecht & Pastashnick Co. Inc., 8 Corporate Circle, Albany, NY 12203, USA, (518) 452 0065

Saab Scania Ab, Gelbgjutargarten 2, Fack 581-01–oe, Linkoping, Sweden

Sankyo Dengyo Co., Ltd 8–11 Chuo-cho 1–chome, Meguro–ku, Tokyo 152, Japan, (03 714 6655)

Sartorius Instruments Inc., McGaw Park, IL, USA, (708) 578 4298

Sartorius Instruments Ltd, 18 Avenue Road, Belmont, Surrey, UK

Sartorius–Werke GmbH, Postfach 19, D-3400 Gottingen, Germany, (0551) 308-1

Schaar & Company, Chicago, IL, USA

Schaeffer, K., Sprendlingen, Germany

Science Spectrum, 1216 State Street, PO Box 3003, Santa Barbara, CA, USA

Seishin Enterprise Co. Ltd, Nippon Brunswick Bldg., 5-27-7 Sendagaya, Shibuya-ku, Tokyo 151, Japan, (03 350 5771)

Seishin, Environmental Control International, 409 Washington Ave., PO Box 10126, Baltimore, MD 21204, (301) 296 7859

Sepor, Inc., PO Box 578, 718 N. Fries, Wilmington, CA 90748, USA, (213) 830 6601

Seragen Diagnostics Inc., PO Box 1210, Indianopolis, IN 46206, USA, (317) 266 2955

Shapespeare Corporation, 901 Park Place. Iowa City, IA 52240, USA, (319) 351 3736

Sharples Centrifuges Ltd, Camberley, Surrey, UK

Shimadzu Scientific Scientific Instruments Inc., 7102 Riverwood Dr., Columbia, MD 21046, USA, (410) 381 1222

Siemens Ltd, Siemens House, Windmill Rd, Sunbury-on-Thames, Middlesex, TW16 7HS, UK

Signet Diagnostic Corporation, 3931 R.C.A. Blvd., Suite 3122, Palm Beach Gardens, FL 33410, USA, (407) 625 3999

Sigrist Photometer Ltd, 1 Pembroke Avenue, Waterbeach, Cambridge CB5 9QR, UK, 01223 860595

Simon Henry Ltd, Special Products Div., PO Box 31, Stockport, Cheshire, UK

Simonacco Ltd, Durranhill Trading Estate, Carlysle, Cumbria, UK

Skill Controls Ltd, Greenhey Place, East Gillibrands, Skelmerdale, Lancs WN8 9SB, UK, (01695) 23671

Société Française d'Instruments de Contrôle et d'Analyses, Le Mesnil, Saint Denise, France

Sondes Place Research Institute, Dorking, Surrey, UK

Sonics & Materials Inc., Kenosia Ave., Danbury, CT 06810, USA, (203) 744 4400

Sony Image Analysis Systems, 3 Paragon Drive, Montvale, NJ 07645, USA (201) 930 7098

Spatial Data Systems Inc., Galeta, CA, USA

Specfield Ltd, 1A Jennings Bldg., Thames Ave., Windsor, Berks, UK

Spectrex Corp., 3594 Haven Avenue, Redwood City, CA.94063, USA, (415) 365 6567

Spectrex, 13 West Drive, Wethersfield, Braintree, Essex, CM7 4BT, UK

Spectron Instruments Ltd, 38 Nuffield Way, Ashville Trading Estate, Abington, Oxon OX14 1TD, UK

Status [see Faley]

Stauffer Chemical Corporation, Westport, CT 06880, USA

Ströhlein GmbH & Co., Girmeskreuzsrasse 55, Postfach 1460, D-4044 Kaarst 1, Germany, (02101) 606-124

Supaflo Technologies Pty Inc, Unit A, 5 Skyline Place, French Forest NSW 20086, Australia, 61 2 975 6060

Surface Measurement Systems Ltd, PO Box 1933, Marlow, Bucks. SL7 3TS, UK, (01628) 476 167

Sympatec GmbH, System-Partikel-Technik, Asseweg 18, D-3346 Remlingen, uber Wolfenbuttel, Germany, (05336 446 448)

Sympatec Inc., Princeton Service Center, 3490 US Route 1, Princeton, NJ 08540, USA, (609) 734 0404

Systems & Components Ltd, Broadway, Market Lavington, Devizes, Wiltshire, UK

Techecology Inc., Sunnyvale, CA, USA

Techmation Ltd, 58 Edgeware Way, Edgeware, Middlesex, UK

Technotest, 65 rue Marius Auffan, 94300, Levallois, France

Telefunken A.E.G., 79 Ulm, Elizabethstrasse 3, Germany

TEM Sales Ltd, Gatwick Road, Crawley, Sussex, UK

Thermal Control Co. Ltd, 138 Old Shoreham Rd., Hove, Sussex, UK

Thermo-Systems Inc (TSI), 500 Cardigan Road, PO Box 43394, St Paul MN 55164, USA, (612) 483 0900

3M Company, Commercial Chemicals Div., 3M Center, St Paul, MN, USA

Toa Electric Co., Kobe, Japan

Touzart et Matignon, 3 rue Arnyot, 75005, Paris, France

Tracor Northern, 2551 W. Beltline Highway, Middleton,. WI 53562-2697, USA, (608) 831 6511

Trapelo Div., LFE Corp., 1601 Trapelo Rd., Waltham, MA 02154, USA

2000 Inc., Box 20769, Atlanta, GA 33010, USA

Tyler, W.S., Inc., E Hwy 12, Benson, MN 56215, USA, (216) 974 1047

UCC International, PO Box 3, Thetford, Norfolk IP24 3RT, UK (01842) 754251

Ultrasonics Ltd, Otley Road, Bradford, Yorkshire, UK

Unico Environ. Instrum. Inc., PO Box 590, Fall River, MA, USA

UVP, Inc., 5100 Walnut Grove Ave., PO Box 1501, San Gabriel, CA 91778, USA

Val-Dell Company, 1339E Township Line Road, Norristown, PA 19403, USA
VEB Transformratoren und Rontegemwerk, 48 Overbeckstasse, 8030 Dresden,
    Germany
Veco N.V. Zeefplatenfabrik, Eerbeck (Veluive), The Netherlands
Vickers Instruments Ltd, Haxby Road, York YO3 7DS, UK, (01904) 31351
Vista Scientific Corp., 85 Industrial Dr., Ivyland, PA 18974, USA, (215) 322
    2255
Vorti–Siv, 36135 Salem Grange Road, Salem, OH 44460, USA, (216) 332 4958

Walther and Company, Aktiengesellschaft, 5 Köln-Dellbruck, Germany
Warmain International Pty, Melbourne, Victoria, Australia
Warmain, Simon Ltd, Halifax Road, Todmorton, W. Yorkshire, UK
Watson W. and Sons Ltd, Barnet, Herts., UK
Weathes Measure Corp., PO Box 41257, Sacramento, CA 95841, USA
Wessex Electronics Ltd, Stoves Trading Estate, Yate, Bristol BS17 5QP, UK
Wild-Heerbrug Ltd, CH-9435, Heerbrug, Switzerland
Wilson Products Division, ESB Inc., PO Box 622, Reading, PA 19603, USA
Wyatt Technology Corp., 802 East Cota Street, Santa Barbara, CA 93103, USA,
    (805) 965 4898

Zeiss, Carl Inc., One Zeiss Drive, Thornwood, NY 10594, USA, (914) 747 1800
Zeiss; Carl, 7082 Oberkochen, Postfach 35/36, Germany, (07364 20 3288)
Zimney Corporation, Monrovia, California, USA

Author index

Subject index

*The numbers shown in **bold** indicate that a section about the subject commences on that page.*